OF ENAMELING *(continued in back of book)*

W9-DJH-874

1st c. A.D.	2d c. A.D.	3d c. A.D.	4th c. A.D.	5th c. A.D.	6th c. A.D.	7th c. A.D.
					Later Celts and Saxons poured molten glass until 9th century	
		Early use of cloisonné enamel to 8th century				
				Enameled jewelry found in England dated 449 – 450		37 Merovingian coins with enamel buried with St. Cuthbert
					First Byzantine enamels more archaic in character	

The Enamelist

To Arlene Minders
with kindest regards from the author
Sincerely
Kenneth F. Bates

Vulture headdress from composite inlay figure of a goddess, Egypt, Dynasty XXX (380-343 B.C.), height 1¼ inches, width 1⅛ inches. One of the earliest —and extremely rare—examples of enameling (the blue and white scalelike sections in front and back of the headdress are fused glass, i.e., enamel). The gold cloisons are exceptionally fine and minute. (Courtesy The Cleveland Museum of Art, John Huntington Collection)

The Enamelist

KENNETH F. BATES

THE WORLD PUBLISHING COMPANY
CLEVELAND AND NEW YORK

Published by The World Publishing Company
2231 West 110th Street, Cleveland, Ohio 44102

Published simultaneously in Canada by Nelson, Foster & Scott Ltd.

First Printing 1967

Copyright © 1967 by Kenneth F. Bates

All rights reserved. No part of this book may be reproduced in any form without written permission from the publisher, except for brief passages included in a review appearing in a newspaper or magazine.

Library of Congress Catalog Card Number: 67–13833

Printed in the United States of America

To

Katharine, Cornelia, and Benham

CONTENTS

FOREWORD

My colleague, Kenneth F. Bates, claims I do not like enamels; I insist that I do, but with some reservations. Such reservations stem from the differences between professional enameling, much of which is excellent, and amateur work. There is a plethora of the latter, manufactured in near assembly-line quantities, and lacking in style, technique, or originality.

Vitreous enamel has a long heritage and tradition. It is a most demanding medium. If the ingredients are not as precious as gold and silver, still the combination of craftsmanship and small scale, along with the intrinsic beauty of enamel itself, have identified enamels historically as precious materials. This is not to say that enameling cannot be practiced on a larger scale; technology has removed that barrier. Nor must it remain a medium for baubles for the idle affluent. Anyone working with enamel must continue to recognize all the inherent characteristics of the medium.

As a painter who has used wax encaustic for almost a decade, I feel great sympathy for a medium that takes advantage of the application of heat to create textures and color patterns; a medium that suggests fusion and yet, at the same time, must be controlled. Control is the key to enameling, and it comes only through complete knowledge of the medium and its limitations. There is little, if any, room for the "happy accident." Much that is called enameling today does not reflect the characteristics of the medium and that in itself would be reason enough for this book to exist.

Kenneth Bates has great experience as both designer-craftsman and teacher and he has always been generous with his information. For almost forty years he has maintained an outstanding record as teacher and exhibitor and, more recently, as an author.

Despite his generosity, there is one ingredient that he can neither supply nor describe. That is the innate, artistic ability which exists in each of us in little or large amounts. On that point the author and his book should not be misunderstood. He cannot make an artist; he can only guide one.

The responsibility of authorship of a book on any craft is rather a two-edged sword. Here is the professional who wishes to make information available both to his peers and to the serious students aspiring to professionalism. This information is also available to those with unparalleled amounts of leisure time who desire to do something with it. Man has always made things with his hands, but there is a great difference between a strong feeling about what one is doing and a casual amusement connected with leisure.

This volume is an unusual combination of history, craft, teaching, and source of inspiration. These parts represent the author, his interests, and his life in an open and honest way; the way we, his colleagues, see him every day.

JOSEPH MCCULLOUGH
Director, Cleveland Institute of Art

September 15, 1966

INTRODUCTION

Since the publication of my first book on enameling in 1951 (*Enameling: Principles and Practice,* The World Publishing Company, Cleveland and New York), I have felt that a more advanced study of the techniques of enameling should be written.

After about 1930, crafts, including enameling, pottery, weaving, jewelry, silversmithing, glasswork, and welded sculpture, emerged from a standing of secondary importance to one of prime concern in our present culture. Due to the arduous endeavors, convictions, and farsightedness of such organizations as the American Craftsmen's Council in New York, enameling and other crafts have taken on greater significance, having, to some extent metamorphosed from the "do-it-yourself," or hobby stage, to a more professional status. The particular phase of doing craftwork as a substitution for playing bridge or watching TV—or, in fact, of taking evening classes in crafts as a means of escape from the kitchen or the office—has not terminated, nor should it be entirely condemned. However, I do think that the results of the "play-at-it" hobbyist deserve condemnation and that as time goes on our art jurors are attempting to discourage that kind of approach. Certainly one does not rule out the chance piece that might happen to have a kind of art quality that could reveal itself amongst a table of five hundred pieces set before the art jury. Nevertheless, it has always seemed a shame to me that the hobbyist persists with the theory that if he indiscriminately turns out hundreds of pieces perhaps one might be good. Has art not won more consideration than this? Does it not deserve some contemplation or warrant any preconceived planning? Must we base the whole proposition on unfettered emotionalism and hope for the best? Is there no underlying, intrinsic quality that designates a work of art? Assuredly there is, but the difference between the artist and the dilettante is that the artist recognizes this fact and is conscious of its discovery whereas the dilettante does not recognize it unless someone points it out to him. Such a point of view releases the artist from presumably knowing a recipe for art. No artist claims this assumption but he does claim to have imagination, that one facet of the intellect without which he feels life is not worth living. It is not a display of imagination to produce stacks of enameled ash trays without meaning, without taste, or without concern. Alas, this becomes only therapy. It is erroneous for the amateur to allege that he is being "creative" merely because he spends his spare time in the basement workshop turning out a multiplicity of enamels to give to friends. Progress from an empirical point of view is possible but not always conclusive. Enameling warrants some study and tends to become more fascinating as one traces this art and its development historically. This book discusses some of the periods in which enameling has had significance.

It is my intention, also, to extend the basic principles of enameling covered in my previous book to include a more advanced approach for the instructor-craftsman who must give a course in enameling for the first time at the high school, university, or art school level. This I hope to do in a more detailed manner, with emphasis upon contemporary uses of enamel.

Also, I feel that a book suggesting ways of teaching must include some information regarding studio construction and layout, lighting, materials, philosophy, and projects to be attempted. Many teachers have told me of the need for information concerning the planning of an enameling course, the purchasing of kilns and supplies, and the arranging of the classroom. Without such information they have felt insecure about establishing a course. I hope my own experiences will help others.

Seldom does an enameling instructor need to worry about discipline or forcing his students to work; it is often the other way around. He occasionally has to force them to quit their work at the end of the period. Almost anyone finds joy in making something. Even the making of a piece of enamel that the instructor feels is badly conceived has, no doubt, given pleasure to the student who did it. The expression "It was fun doing it" is perhaps the most discouraging for the instructor to hear, especially when he feels that the work is inferior. By "inferior" I do not necessarily mean that the craftsmanship is poor. So often, what is sadly lacking, is a sense of design, a feeling for proportion, appropriateness, or color. Therefore, it seems to me, that any book on enameling (or for that matter, craft books in general) should stress the need of design study. I do not deny that one can be overtaught in design theories, and that one becomes so conscious (or self-conscious) of the do's and don'ts in design that a fresh and spontaneous appeal is overlooked, but so often this is not the case. After all, the alert instructor will spot native ability when he sees it in the classroom. It is for him to convince the students that there are certain design standards, certain criteria about designing that make a thing "feel" right, or "feel" awkward. It is not the intention of this book to presume to tell the artist-craftsman *how* to design, but to show several examples of variations allowing the craftsman's sense of taste to be his own guide through selectivity.

The word "design" is a very broad term, and even though I insist that anyone who cares to study enameling seriously should first have been exposed to a course in basic design, still the problems involved in applied design are those that concern us most. Whether putting a design "on" something becomes an enhancement to the object or whether it is destructive of the object's intrinsic beauty and results in nothing more than superfluous decoration is the enamelist's problem. Technically, enamel designs should look "in" the surface, not "on" it. So, aesthetically, the design must appear to be "for" it, not added "to" it.

We must consider the contemporariness of this design, as against something entirely out of relation to our present-day direction. A recapitulation, a rediscovery, or a re-evaluation of some former style or era has validity only if the artist has sincere convictions and his thinking is the result of considerable research.

The study of color, its properties, its dissonances and harmonies, its illusions and visual possibilities can be studied scientifically, up to a point, but when it comes to personal reactions to color, personal taste in selection of colors there is little that can be taught. It stands to reason that in the case of enameling where color is a matter of serious consideration, an empirical approach is the only one; one must gain information by constantly testing glazes, their reactions to various heats and variable timing.

Perhaps for the enamelist, his conceptual result is on a more precarious footing than for the painter. The painter applies his color directly to the canvas, but after the enamelist has established his color scheme, he still has the added hazard (and often delightful suspense) of fusing it in the kiln. Some of the effects of timing, overfiring, and underfiring, I shall include in this book.

I am invariably concerned with the individual, a concern that in no sense is unique with me as this has often been spoken of as the "age of the individual." However, it appears to me that in the elaborate and involved field of art, any artist can only vouchsafe, or in fact communicate, that which is within himself. All too often we witness those who disseminate, who promulgate that which is quite foreign to themselves. We do not presume to be like Pablo Picasso with such encompassing gifts that nearly any phase of art is within our grasp of creativity, or like Leonardo da Vinci, whose extraordinary mental superiority ascertained his heaven-born genius in many fields. We do, however, have the right to think that, in some small way, each of us has his forte, his interest, his inquisitiveness. With me, it is the wondrous world of flowers and plant life, and I propose to write something about it in this book, something of its magical enchantment, something of its undeniable value to the designer in particular.

This book includes examples of how flowers and plants have been used in a desecrating or degenerate manner, how all of the subtleties and nuances of this fascinating object-subject can be reduced to what is known as "bad taste." It is my sincere hope that those who peruse this book will not turn to such plates first.

Perhaps an artist should have a garden, perhaps he should grow plants and flowers indoors in the winter. He who is concerned with that outmoded, and misused word "beauty" should not deny himself a few potted geraniums, straggly though they may be, or a Dieffenbachia plant, poison to the touch, but refreshing to the eye.

I appreciate the cooperation of the Cleveland Museum of Art; the Metropolitan Museum of Art, New York; the Walters Art Gallery, Baltimore; the Cooper Union Museum, New York; the Carnegie Institute, Pittsburgh; the Museum of Contemporary Crafts, New York; and the American Craftsmen's Council, New York, for the use of photographs in this book. The ink drawings and some of the other photographs were done by the author, but, by far, the most professional photographic studies are the result of long hours of painstaking work by my good friend Donald Schad of Western Reserve University.

I am indebted to and wish to acknowledge my gratitude to three other groups of people—my artist-craftsmen friends, my gardener and floricultural friends, and most sincerely to my young art students. It is they for whom I have the greatest love and in whom I have perennial faith and respect. Without their constant search for knowledge and their accreditation of my attempts to guide and teach them, I could not have written this book.

Kenneth Bates

September 1, 1966

CHAPTER 1

HISTORICAL SIGNIFICANCE OF ENAMELING

It is a common but surprising occurrence to converse with otherwise well-informed people who have no knowledge of enameling, or of its long heritage and historical significance. To millions in our fast-moving twentieth-century culture an enameled ash tray in their homes or offices represents little else than a utilitarian object. It has to them, presumably, no more meaning than a tray made of molded glass or anodized aluminum. When an interest is shown regarding enamels, one is apt to hear this question asked with expected regularity: Is enameling something new, or have people known about it for some time? The fact is that man has been making enamels since the sixth century B.C.

Enameling is an art as old, if not older, than the art of pottery, but for some reason pottery is more widely understood than enameling, and of course it is of considerable archeological importance. Because enameling is not produced on a clay base but on precious metals, and often combined with jewels of great value, throughout history enamels have been stolen, or hoarded, by the great families of the nobility and transported from nation to nation, sometimes being hidden or lost through the centuries.

Enamels of historic value are becoming more and more difficult to purchase. Actually, countries have few ancient enamels to offer to the collector, even though gold and enamel are veritably nonperishable. The exquisite state of preservation of the Byzantine enamels in the Dupont collection at the Metropolitan Museum of Art, or the brilliance of color seen in the fabulous Guelph Treasure pieces in the Cleveland Museum of Art, proves the fact that historic enamels have not crumbled into dust, but, on the other hand, are so treasured by individuals that public museums have simply not been able to locate many of them. We cannot escape the romantic, the historiographical nature of enamels. The translucent cobalt blue enamel used by students and professional enamelists today is identical with the cobalt blue found in the background of Léonard Limosin's portrait of François de Cleves, who was one of the famous enamelers of the Limoges school in France during the sixteenth century.

Perhaps the beginning student or the hobbyists should be more aware of the background and permanency of enamels, which would, in turn, create a greater respect for the medium. In past centuries the enamels were thought of more as a precious stone or jewel and often used to refurbish the house of God. Their depth of color and limpidity of tone evoked mystical connotations. Enameling was constantly used as an accompaniment to gold for tiaras, neck bands, large costume clips and brooches, especially during the Gothic and Renaissance periods. In other periods, namely the great Byzantine peak of the tenth century, the German Mosan school of the twelfth century, and the famous Limoges, France, period during the middle of the sixteenth century, enameling, like sculpture and mosaics, was a very substantial form of communication. The stories of the Old and New Testament were graphically depicted in enamels for the benefit of the masses of illiterate peasants.

A certain amount of historical data is extant regarding enameling, however, there are still several periods in enamel chronology about which little is known. We can only hope that as archeology continues to search and probe with recently discovered scientific methods, the mystery of lost enamels will be unearthed.

1. *Greek*

Our survey takes us back to the Greek civilization in about the sixth century B.C. where the oldest enameling is assumed to have been fabricated in the form of encrustation on small figurines. This appears to be some type of vitreous substance, and because it is fused to a metal, usually gold jewelry, we may call it enameling. According to a documented evidence, wires were used to separate a kind of matte finish glaze that appears at a lower level than the wires; we can therefore safely say that the oldest known type of enameling is cloisonné. Exportation of Greek enamels to Russia and Asia Minor has since been discovered. In the fifth century B.C. some of the Greek sculpture showed areas of gold with enamel applied. This effect often occurred on parts of drapery and embellishments to the costume. There are several examples of ancient Greek goldsmith work showing small inlaid motifs between gold fences or cloisons. There is little doubt that these pieces of great antiquity represent the true type of cloisonné enameling as it is known today.

2. *Celtic and Roman*

Throughout western and northern Europe one is constantly reminded of the vestiges of a noble and permeating civilization, known as Celtic. Particularly in northern Spain peasants still carry on many of the traditions left by the Celts. There, in the former province of Galicia, farmers today reap gorse, a low, straggly, yellow, blooming shrub, in much the same way as the ancient Celts did, using oxen and crude implements. They speak in a Celtic tongue, build fanciful chapels above the granaries, and take enormous pride in their Celtic background. The Celts had already reached western Europe in 2000 B.C. By 1000 B.C. they had entered the Low Countries along with the Teutonic tribes. The Gauls, a Celtic tribe, were occupying much of France in the seventh century B.C. and by 600 B.C. had invaded across the Pyrennes. Celts in 700 B.C., with bronze already in use, were in Britain, and later

FIGURE 1. Cup, Provincial Roman, second half of second century; champlevé enamel on bronze, silver lining and base. (*The Metropolitan Museum of Art, Fletcher Fund, 1947*)

in about 400 B.C. a tribe known as the Brythonic Celts invaded England and Wales, forcing other tribes such as the Caledonians into Scotland.

We attribute the earliest records of a type of enameling known as champlevé to the Celts. They had found a way of pouring molten glazes into depressions and decorative areas that had been carved on their swords and shields. These vitreous substances hardened when cool and evidently satisfied the aesthetic tastes of these early warring tribesmen.

Before the year 50 B.C. the Romans had reached the river Rhine, Great Britain, France, and later in 19 B.C., Spain. Artifacts of Roman origin and Celtic origin are often very similar in nature and consequently confused. A majority of this early work was done on bronze, but customarily in the champlevé manner.

3. *Egyptian*

The first pyramid was built in Egypt in 3000 B.C. and to date, notwithstanding the magnificent artifacts, sculpture, costume embellishments, and sarcophagi paintings, there is little proof that enameling per se was accomplished in Egypt until Rome was founded in 750 B.C. The fact remains that Egyptians did make use of colored glass and vitreous glazes on pottery at a very early period in their civilization. There have also been found examples of bits of colored glass that were *glued* not *fused* to a metal or clay form. One technique that we often read about, called "millefiori," may have been confused with enameling. Millefiori is a method of decoration whereby tiny rods of colored glass are fused together and then cut across to achieve a very rich and colorful effect. The same technique was used later in the Gothic and Renaissance periods. Caucasian enamels, sometimes confused with those of Egyptian origin, have been recorded as early as the first century B.C., this being what is known as the early part of the Iron Age. These enamels have a certain relation to the Greek and Saxon enamels, and even though the similarity is distinctly Egyptian in character, still the true authenticity is unknown. So far Egyptologists have not found enough proof to support a claim that enameling was one of the better known crafts of that amazing civilization—at least, not in the periods we are accustomed to think of Egypt. A rare example of Egyptian enamel is shown as the frontispiece to this book.

4. *Japanese*

A considerable number of early Japanese enamels made with fine gold wire and consisting mainly of designs composed of flowers, dragons, and clouds have been authentically dated as early as the third to the sixth century A.D. We can relate this period to Mohammed's birth in 570 and his publishing of the Koran in 612, which should impress the layman of the ancientness of the art of enameling. Japanese enamels have always been of a meticulous nature, reflecting the patience and assiduousness of the Oriental mind, which is well trained to meet the

FIGURE 2. Vase, Japanese, nineteenth century; cloisonné enamel. (*The Metropolitan Museum of Art, bequest of Stephen Whitney Phoenix, 1881*)

demands of the medium. Later, in the eighth century, archeologists discovered that a few enamels were made—in particular, a mirror back, but through the ages this piece has suffered disrepair, and was, even in its original state, a rather primitive piece. Japanese craftsmen have continued to make cloisonné enamels through the nineteenth century. During the late nineteenth and twentieth centuries the Japanese have developed and adapted other methods that are somewhat in competition with the French enamelers. Their metal foils and brilliant colors have little of the soft dull-finishes found in the work of their early progenitors. During the sixteenth century, however, much of the Japanese cloisonné work was copied profusely from the Chinese; many museums hold exquisite examples of their fine craftsmanship. More modern Japanese enamels employ designs of a less artistic nature. Naturalistic paintings of birds, chrysanthemums, dragons, and symbols have deteriorated into a representational style, often imitating water colors. Still impressive, however, are the ultimate skill and the inconceivable patience of the Japanese craftsman. Timewise, even to approach the American wage scale, any one of these Oriental enamels would have to be valued at thousands of dollars.

5. *Anglo-Saxon*

In the year 300 A.D. raids into Great Britain were made by the Angles and the Saxons from Germany. By the fifth century the Anglo-Saxons had consolidated themselves in South East England. Later in the same century we have the romantic story of King Arthur, who, after numerous predatory incursions, harried the invaders, who eventually moved to Scotland. This entire period in English history is one of the most romantic and enchanting eras for study. Significantly, enamel work was one of the prevalent forms of decoration and embellishment, particularly on bold clasps, buttons, belts, and brooches. A white enameled brooch, in excellent preservation, on exhibit in England, is dated about 590 A.D., a time when most of England was in Anglo-Saxon hands—a period otherwise known as the "Dark Ages." A most famous cloisonné enamel called the "Alfred Jewel" was made for Alfred the Great (849–901). It has cells of opaque white enamel, surrounded by green flowers in a most bold and intriguing design and is unmistakably Anglo-Saxon in character. The setting in red gold also shows an intricate, pebbly surface, which was made by fusing minute balls of gold onto the metal. "Granulation" is the name given to this kind of work. It is extremely difficult, and few modern craftsmen practice the art but the Anglo-Saxons made extensive use of it.

I shall describe only one other enameled piece from this intriguing period. It is a purse-mount from the *Sutton Hoo* ship, part of a burial treasure excavated in Southern England. Being of solid gold, the purse, with mechanically perfect hinges, is in excellent condition. There are two hexagonal plaques with the most delicate cloisonné enamel work. So often in this period garnets were used, many times for personal adornment. The color of garnets seems to be an appropriate selection considering the rough textured costumes, leather jerkins, and flowing capes worn by courageous knights. In the above mentioned purse-mount there appear one hundred minute cells, or bezels, to receive the garnets. The enamel is of a garnet color also. Ducks, falcons, and armored warriors complete the design in a cloisonné technique called "lidded cloisonné"; that is, a kind of fitted cell with thick gold lids. There were some 37 Merovingian coins within this purse when it was discovered. Other artifacts showed the profusion of garnet stones set over gold foil for added brilliance and subsequently set in a gold framework.

This same technique and deep red color emphasis were found on a magnificent cross buried with St. Cuthbert, who died in 687 A.D.

In a later Saxon period (tenth century) there were produced other and even finer examples of enameling work. In the British Museum a gold brooch about 1½ inches in diameter consists of a convex circular disc of cloisonné enamel, also with filigree and granulation. The enamel, which is blue, green, yellow, and white, is of good quality and perhaps in better repair than that of the famous King Alfred jewel. On the face is a figure of a king, full front view, in extraordinary graphic delineation. This piece has aroused some question amongst historians, as it connotes certain Byzantine characteristics. However, the flesh colors used in this brooch are not the same as the Byzantine colors, and it is thought by some to be of Germanic or English origin modified by the enamelers of Lombardy in later centuries. Nevertheless, this tenth century brooch is a rare and important jewel historically.

6. *Indian*

In our mention of India as a source of historic enamel research we must go back a bit chronologically to mention a pendant of the early part of the fifth century. This is a beautiful example of cloisonné enamel from Sirkap and now a part of the Cleveland Museum of Art collection. This important piece of Indian jewelry from Sirkap (Northwest India) is in a classical Greek style, and on first glance it appears to be made of gold with an enamel inset of deep red. However, upon close scrutiny one must agree with the historians and admit that what resembles enamel is actually thin slices of carnelian, and thus the jewel does not belong in our classification of Indian enamels. Its mention is justified, however, as it leads to a discussion of the amazing transparent ruby red enamel that it emulates. Nowhere has such a brilliant red been achieved. In Jaipur, where the best work in India was done, the red was always in excellent harmony with jewels. In fact, the Indians who always worked on gold were known for their fiery red. It has been said that "the ability of producing a good transparent red is the measure of the enamelists' talent."[1] Labarte in his *Handbook of the Arts of the Middle Ages* says that enameling originated in Phoenicia,

[1] S. S. Jacob and T. H. Hendley, *Jeypore Enamels* (London: Elm House, 1886).

FIGURE 3. Lota (covered jar), Indian, Jaipur, ca. 1700; cloisonné enamel on gold. (*The Cleveland Museum of Art, J. H. Wade Fund*)

then found its way to Persia in 531 to 579 A.D. From there it was carried on by the Greeks and then by the Indians.

In any case, a great deal of enameling was done in India for the Maharajas. The most ancient piece of Indian enamel was accomplished by a staff of craftsmen working for the Maharaja Man Singh of Jaipur. This fantastic piece is composed of 32 cylinders, each enameled with figures, flowers, animals, and landscapes. The old techniques were slow and laborious. This piece has taken 10 firings for each segment, which represented 3,000 rupees worth of gold and took 6 years to complete. Each workman did a certain kind of work (sitting in the Oriental fashion on the floor with legs crossed). One worker prepared the metal, another placed the enamels on the jewel, another did the firing, and still another the polishing. Each enamel was applied according to hardness in the following order: white, blue, black, yellow, pink, green, and red. The pure ruby red was the most fugitive and only the Jaipur artists gave the true gold luster through the red. (The gold was pure except for $1/96$ part copper.) Evenness of application, enamel ground to a paste in a mortar and pestle, and acids made from fruits were the Jaipur secrets.

Some of their varied uses for enamel were jewelry (always with enamel on back and front and between the jewels), anklets, bangles, handles of swords and daggers, umbrellas, fans, camel trappings, sleeve links, card cases, match boxes, whistles, and even a small compass in a mango-shaped locket to indicate the direction to Mecca.

Modern Indian enamel is less artistic than the old because of the speed in its production and the need for quick profit. It would be a sad

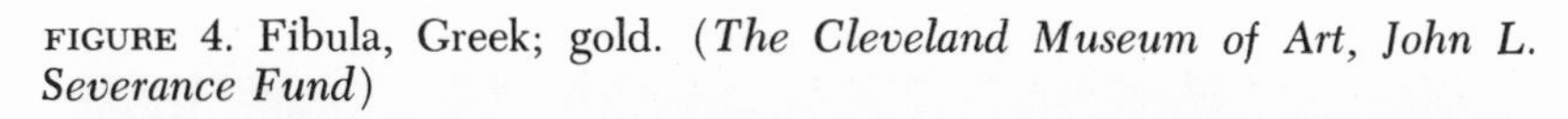

FIGURE 4. Fibula, Greek; gold. (*The Cleveland Museum of Art, John L. Severance Fund*)

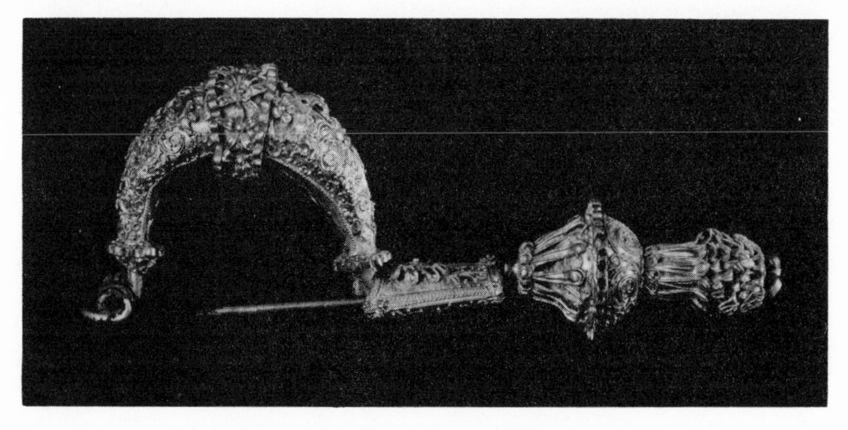

mistake for anyone who has become interested in enamels not to view the magnificent covered jars, pendants, lockets, and amulets owned by the Cleveland Museum of Art, some dating from 50 B.C. and others as late as the eighteenth and nineteenth century.

India and Persia produced considerable enamel work in the sixteenth century. This was often of the cloisonné type and of amazing intricacy and elaboration. Involved scrolls, complicated filigree, and superimposed elaboration marked the Indian and Persian enamel work. A dagger of Shah Gahan and other festival daggers exhibit extraordinary skill in the technique. The present-day Sikhs are still in possession of many of these valuable pieces. Enameling was an art that attracted adventurers, pirates, and visitors. Spreading to Delhi, Jodhpur, Benares, and other cities, the Indian enamels that were once thought of as extremely luxurious articles eventually became more common. There were scent lockets, plates, bowls, charms, small and large vases of every description, which soon changed from unique pieces to objects of multiple duplicity. The Indian and Persian craftsmen were skilled at even another phase of enamel work. This consisted of engraving the surface, a process we now call repoussé. By hammering the metal both from above and below, a three-dimensional, or modeled, look was given to the surface. The enamel was subsequently inlaid in given areas to produce a sense of richness. Their use of blue, green, and opaque white, combined with black enamel, gave a vibrating effect, but because of the desire to produce in large quantities for saleability, the work was often crude. At the present time Indian enamels, usually done on bronze or brass, are occasionally seen in gift shops in the form of cigarette cases, boxes, and ash trays. These have a decided and unmistakably commercial look about them, reflecting little of the finer and more traditional Indian enamel work. In New York City I was able to purchase a small locket, Indian in character and authenticity, which represented genuine East Indian craftsmanship as we prefer to think of it. Most intriguing lacery of pure reddish gold was exquisitely encrusted with sparkling ruby red translucent enamel, supplemented by minute areas of opaque white having details and refinements exploited by the use of hair line black overglaze.

7. *Byzantine*

My purpose in mentioning Byzantine enamels here is mainly to record this important episode chronologically. So much has been written about Byzantine enamels, both pro and con, that the student of enamel history should have no trouble in finding a wealth of material. I am sure that to add more superlatives regarding this magnificent period would only become redundant.

At the beginning of the sixth century, the early Christians, who had formed an empire in Byzantium some two hundred years before, began to make use of molten glass upon precious metals. Their style of architecture had already been established and consisted of rounded arches resting upon columns, which formed a domelike structure. The vaulting that served to connect the area enclosed by both arches and domes was known as a pendentive. Such complex and advanced architectural structures deserved refinement and forms of embellish-

FIGURE 5. St. George medallion, Byzantine, eleventh century; one of a set that was on an ikon of St. Gabriel formerly in the old church of the monastery at Jumati in Georgia; cloisonné enamel. (*The Metropolitan Museum of Art, gift of J. Pierpont Morgan*)

ment that would represent man's highest spiritual achievement for the house of God. Mosaics was one of these forms developed to a state of grandiloquence for interior wall surfaces, and gold cloisonné enamel became the other, being used principally for objects within the basilicas.

Although the first Byzantine enamels were made in the sixth century, the more rare examples often brought back by the Crusaders at a later date were done from the eighth to the eleventh century. The peak of the Byzantine work was during the tenth century. By some aspirational urge, some spiritual apprehensiveness, the enamelers of that time elevated their talents from the commonplace, achieving unprecedented esthetic results. It is safe to say that up to the present time no enamels have been more subtly designed, more delightfully stylistic or handsomely colored than those coming from the hands of Byzantine craftsmen during its most flourishing period.

Byzantine enamels are almost always done in the cloisonné manner and on pure gold with very fine gold wires. Their colors, both opaque and transparent, are well defined, clear and simple, producing a strong, bold, but always sensitive design statement. One of the techniques used (one that might present cost problems for the modern craftsman) was to employ a full sheet of hammered gold about 22 gauge thick, sink or set back the areas to receive the enamel, and within these sunken areas make use of the gold cloisons for the definition of the design. Many such enamels, seldom other than religious in nature, can be seen at the Metropolitan Museum of Art in New York. There are collections of Byzantine enamels all over the world. Hieratic figures, dancers, as well as royally and elegantly garbed characters, were used as models for enamel work. If, according to historic document, enameling progressed from Egypt, Asia Minor, Russia or Persia to Byzantium, it is clearly ascertained that it spread rapidly, achieving immense popularity from Byzantium to Spain, Italy, England, and Germany.

8. *Chinese*

So often when one thinks of China's ancient civilizations one is led to believe that all forms of art expression existed there prior to other countries. Although the first Chinese dynasty was established as far back as 2200 B.C., the Great Wall of China was not started until 250 B.C. As for any trace of enameling, it wasn't until the fourteenth century when China had made contact with the West that cloisonné was done. There is some record of a kind of work that was first thought to be cloisonné enameling and dated about the first half of the seventh century, but this has been proven to be a type of craft made by cementing small pieces of glass into what simulated wire cloissons. (Early Egyptian work presents the same kind of confusion.) The first dated piece of Chinese enameling, a small cup with dark green and turquoise coloring was done in 1426. The motifs represented were peonies, chrysanthemums, clouds, and scrolls in a rambling floral pattern done on gilt, a metal composed of bronze, brass, tin or copper, having a color quite similar to gold.

From the second half of the fifteenth century many incense burners, boxes, jars, and other artifacts have been preserved. Nearer the end of the fifteenth century cloisonné, which was the most popular form of all Oriental enameling, was becoming more elaborate and showed the use of mixed colors. Often these enamels became quite realistic, with careful shadings and details. Such animals as ducks, dragons, apes, and tigers were more like porcelain painting than enamels.

FIGURE 6. Container in form of unicorn, Chinese, Chien Lung; cloisonné enamel and gilt copper. (*The Cleveland Museum of Art, The Norweb Collection*)

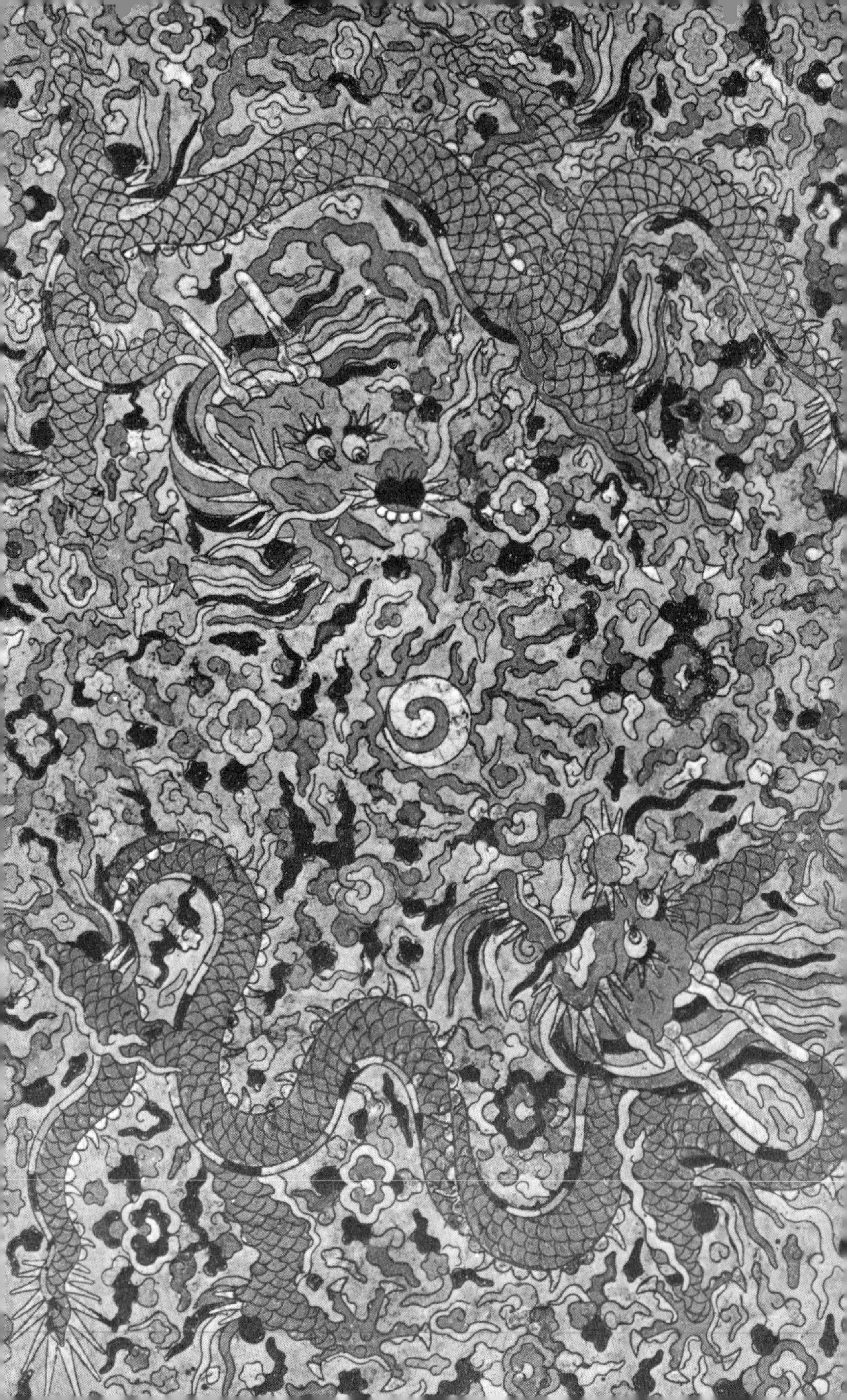

After the Ming dynasty (which extended from 1368 to 1644) until 1722, some of China's best enamels were produced. Colors improved, formulas were found for grinding various oxides of metal with the silicates, shading and drawing became more refined, and the beautiful soft finish typical of Chinese enamels was at its highest state of perfection. There were numerous factories with trained workers producing quantities of objects for export. Buddhist temples were replete with statues, reliquaries, vases, and sacred objects all done in a most laborious and painstaking manner. Myriads of fine gold wires separated the tiny accents of color and literally covered the surface.

At the present time Chinese enamels are a common commodity in many gift shops in America. These in no way resemble the fine, old, and more traditional enamels. Often the background is black, and although the motifs are Chinese in character, the colors have lost their charm, being more primary and crude. Since painted enamel was introduced to China from Limoges, France, in about 1735, much of the cloisonné technique was forfeited and in its place the faster and more expedient painted enamels were popularized. Now a small tray of Chinese origin is not difficult to find in almost any tourist shop priced at one or two dollars.

9. *German*

Enamels in Germany, especially those that originated along the valleys of the Rhine and the Meuse Rivers in the eleventh and twelfth centuries, are of outstanding significance to the collector and the historian. Here a bold and simple style of great charm and naïveté, with an almost primitive quality, typified plaques, crosses, altar pieces, and religious work in general. Godefroid de Claire attained an enviable reputation as an enameler in the twelfth century. Figures were rendered flat with arbitrary shadings. Outlines were severe, but strangely expressive. Most of the work was done in champlevé, with occasional areas of cloisonné, and the addition of carbuncles and precious stones in crude bezel settings. The Cleveland Museum of Art is the proud possessor of a very famous collection consisting of altar pieces, crosses, and reliquaries, which through the centuries has come to be known as the Guelph Treasure.

The school of enamelists from the Meuse Valley, called the Mosan school, is significant. It produced many fine pieces during the eleventh and twelfth centuries. One cloisonné enameled brooch depicting a female figure with necklace recalls the jewels of Empress Theodora

FIGURE 7. Panels: dragons guarding jewel, Chinese, Ming Dynasty (1368–1643); cloisonné enamel on copper. (*The Metropolitan Museum of Art, gift of Edward G. Kennedy, 1929*)

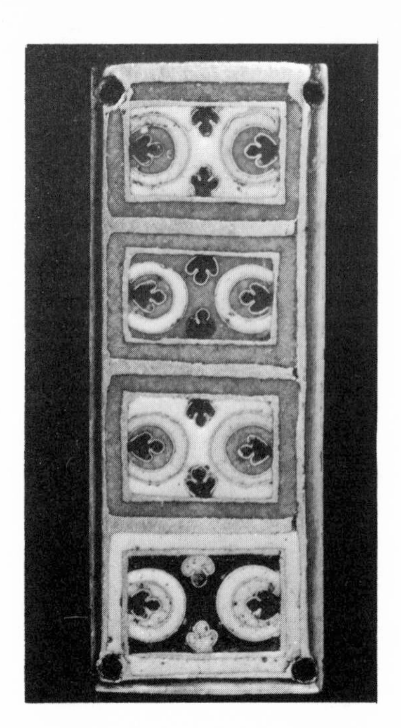

FIGURE 8. Plaque, German, Cologne, ca. 1170; champlevé enamel on gilt copper. (*The Cleveland Museum of Art, purchased from the J. H. Wade Fund*)

FIGURE 9. The Gertrudis Portable Altar, German, Brunswick, Lower Saxony, mid-eleventh century; oak wood casket, porphyry slab, covered with gold in filigree and embossed work, niello, cloisonné enamel and precious stones. (*The Cleveland Museum of Art, John Huntington Collection*)

seen in the Ravenna mosaics. The Byzantine influence is apparent but such influence often permeated all of Europe and has been a constant confusion to many historians.

We cannot escape the foreseeable relationship between enamel work and that of the fine illuminated manuscripts. The same style of elaborated surfaces and of decorative devices (scrolls, floral patterns, birds, vines, etc.) were used by the manuscript writer and the enameler of that time. One influenced the other during the centuries leading up to the Middle Ages.

FIGURE 10. The Gertrudis Portable Altar (detail), German, Brunswick, Lower Saxony, mid-eleventh century; oak wood casket, porphyry slab, covered with gold in filigree and embossed work, niello, cloisonné enamel, and precious stones. (*The Cleveland Museum of Art, John Huntington Collection*)

10. *Middle Ages*

It is permissible to call the period from 1100 to approximately 1300 the Middle Ages, at least the "Early Middle Ages." The period from the year 1477 to about 1506 we will speak of later and use the term "Late Middle Ages." These terms may be somewhat ambiguous but in order to pinpoint the chronological progress of the growth of the enameling art in Europe they are justifiable. There are few enamels discovered in England that were made during the seventh century or directly after the Anglo-Saxon invasions. However, in the following period, the twelfth to fourteenth centuries, we find that filigree work and enameling held the field in a variety of forms. Enameling flourished in Scotland and was a still more vital form of art expression in Ireland coincidental with the time of the Norman Conquest, 1066, and until

FIGURE 11. Reliquary cross, German, twelfth century; cloisonné enamel. (*Walters Art Gallery, Baltimore*)

1172 when Henry II conquered Ireland. In 1314 England and surrounding countries were enduring the horrendous plague known as the Black Death, and it seems rather ludicrous to be recording the continuance of a craft as time-consuming or as painstaking as enameling during the period when the image of death was at every door. However, the nobility still commissioned artisans of repute to perform miracles, to produce objects of personal adornment, and gifts for their ladies of rank. An early thirteenth century crown for the reliquary of St. Oswald at Hildesheim is of exceeding interest, made of gold and set with innumerable plaques of enamel and semi-precious stones. Edward I, who died in 1290, had belts of gold and enamel, gold pendants, gold rings, brooches, and gloves all beautifully set with enamels, often in bold quatrefoil designs. In this period again we find that the manner of the illuminated manuscripts and also the stained glass window designs dominated the thoughts of the enamelist.

11. *Late Middle Ages*

As the Middle Ages progressed to the Late Middle Ages (1477 to 1506) gem cutting became less important and was more or less forsaken for enameled filigree work. This effect was sometimes used in conjunction with marginal designs for illuminated manuscript work.

FIGURE 12. Candelabrum, fourteenth century; rock crystal and champlevé enamel. (*Carnegie Institute, Pittsburgh*)

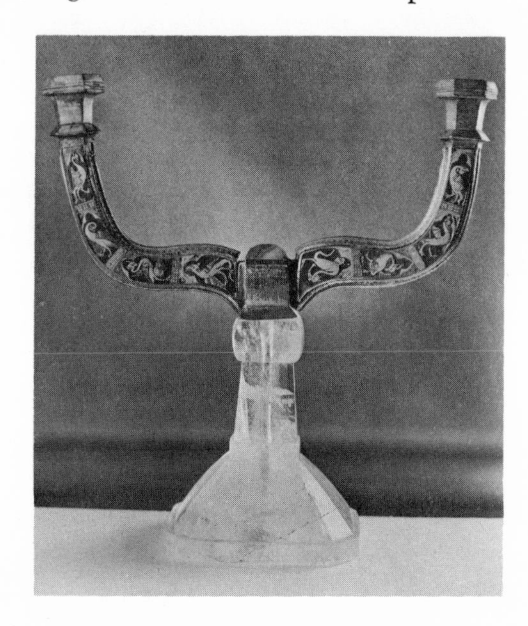

Heraldic decorations employed the use of enameling frequently. Reliquaries were in great demand and Limoges (France) on the pilgrimage route to Santiago de Compostela in Spain did a flourishing business. Now the flamboyant and perpendicular type of architecture that was much in evidence had a definite influence on the enamel designs of the period. Low necklines and crowns for the great ladies of France and England were the fashion in 1455. Margaret of Austria owned a heavy collar or band of gold formed of 27 pieces, each point in the design meticulously enameled in black, white, and light red. Another collar was composed of 43 roses with diamond centers joined

FIGURE 13. Pendant with pelican, German, ca. 1600; enameled gold with rubies. (*The Cleveland Museum of Art, Mr. and Mrs. Severance A. Milliken Collection*)

by little snakes, all enameled with considerable skill. And so we might continue with our descriptions of the fantastic enameled jewels worn by Suzanne de Bourbon, duchess of Savoy, and numerous other elegant ladies of the period. Balls of ambergris and musk within gold and enameled lockets hung from the neck of Charles V. When Henry VIII was married to Catharine of Aragon in 1509, her wedding gifts consisted of hinged plaques, belts, lavalieres, and as a special gift, a complete iconographical triptych in gold and brilliant enamels supported by long rosaries.

12. *Gothic*

The term "Gothic" is ethnical as well as esthetic and architectural. People thought, acted, and worshipped in a manner unlike their predecessors. For example, art gained certain footholds, and in particular, the craftsman became as important in the social structure as certain tradesmen. In the year 1331 the goldsmith craft was officially recog-

FIGURE 14. Girdle, Italian, late fourteenth century; ornaments of silver and basse-taille enamel on silver. (*The Metropolitan Museum of Art, gift of J. Pierpont Morgan, 1917*)

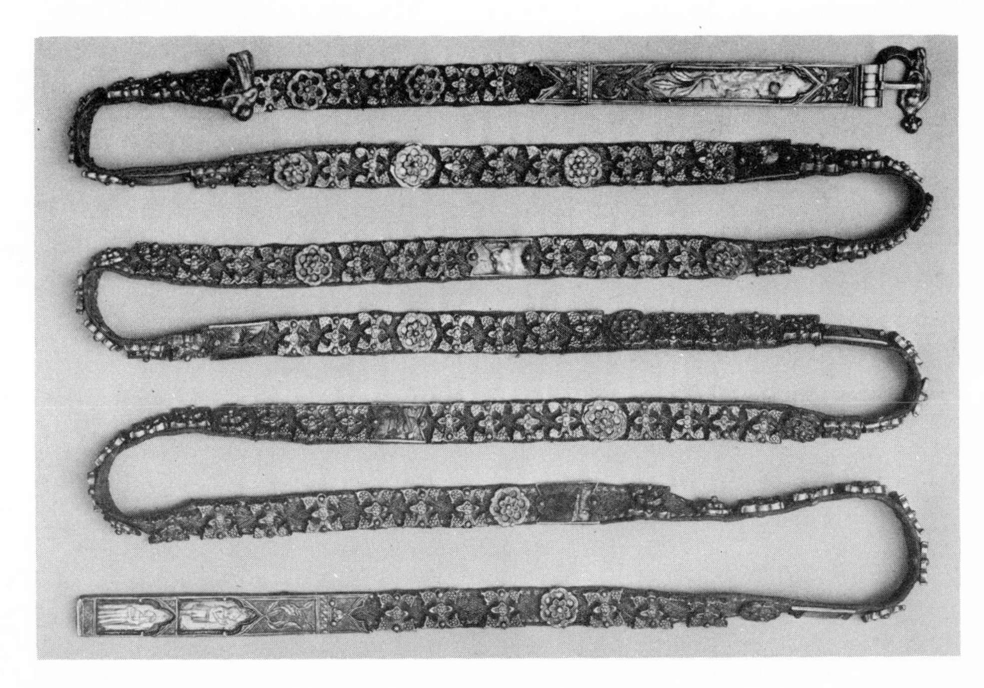

nized by the government. Although luxury was rampant in the French court, the ordinary people were controlled by strict laws forbidding finery of any sort. At the same time, about 1363 Edward III of England decreed that "cloth of gold" be forbidden as wearing apparel or even to be possessed by any common people. But in 1380 the royalty dressed in the most exorbitant finery. Kings were not satisfied with one crown but had fabricated for themselves many crowns of different designs, made of gold and set with precious jewels and exquisite enamels. Ordinary felt hats owned by the upper classes were sewn with threads of gold and set with enamels and jewels. Belts, clasps, rings, and brooches of ostentatious proportions were often enameled. A magnificent wheel brooch with basse-taille or repouséed enameling was produced for Queen Clemence of Hungary in 1328. In the first half of the fifteenth century we find other uses of enamel such as that seen in the famous brooch made for the Duke of Burgundy (now in the Imperial Treasury at Vienna) and also a charming pair of enameled lovers originally owned by the Duchess of Orléans.

13. *French Limoges*

The French have traditionally been known as innovators, and in the history of the art of enameling France has played a very unique and important part. An unknown Frenchman discovered (possibly by accident) in about 1450 that even after subjecting the colored enamel grains to heat hot enough to cause fusibility, the planned areas of the design would not run together. In other words neither wires (cloisons) nor gouged or etched cavities (champlevé) were necessary to separate the colors. Then in Flanders in the middle of the fifteenth century the idea originated that enamel could be superimposed on enamel, a technique unheard of before. Along with this idea came the discovery of retaining the surface enamel to the plate by firing enamel on the back (counter-enamel) to equalize the stress and prevent the surface from cracking off. The idea spread almost immediately; by 1490 a French town in the Southern part of the country by the name of Limoges had become the center for the new technique. The type of enameling, known as Limoges enamel, is more than simply painting the surface in imitation of an oil painting. Many layers of translucent enamel are fixed onto the copper plate, and sometimes applied over layers of opaque white enamel. The Limoges enamels of the fifteenth and sixteenth centuries can be seen in many museums today. The production of this kind of work increased quantitatively if not always

FIGURE 15. Gemellion, French, Limoges, second half of thirteenth century. (*Detroit Institute of Art*)

qualitatively. Secret methods and techniques were inherited within families, and many names like Nardon and Jean Pénicaud, Léonard and Eduard Limosin, and Pierre Raimond are well known today. Portraiture, classical subjects, and religious illustrations became common, the artisans often showing pagan and sensual qualities. In fact, some of the later Limoges pieces, such as a copy of "Passion" by Albrecht Dürer, lack the imagination and mastery of the earlier crafted pieces, and eventually such copying of paintings reduced the famous school of Limoges enameling to a state of deterioration almost equal to that of American China painting.

FIGURE 16. Henry d'Albret; French, Limoges, sixteenth century, Léonard Limosin; painted enamels. (*The Metropolitan Museum of Art, Jules S. Bache Collection, 1949*)

FIGURE 17. Ewer, Italian, Venice, fifteenth-sixteenth century; painted enamel on copper. (*The Cleveland Museum of Art, gift of J. H. Wade*)

FIGURE 18. Two oval plaques, French, eighteenth century (1750–1800); painted enamel on copper, signed Micault. (*The Metropolitan Museum of Art, gift of Mrs. Morris Hawkes, 1925*)

14. *Renaissance*

In the Renaissance period commencing with the sixteenth century and continuing through the seventeenth we find enameling now used to cover partially surfaces of figures in the full round. Basse-taille enameling (that which is modeled and formed from the back) was in vogue, and logically so, as the concept of sculpture and painting began to open new vistas. The giant in the world of craftsmanship at this time was, of course, Benvenuto Cellini. In 1534 he created the famous salt cellar, which is now world renown. During the seventeenth century great enameling work was being done in East Germany, South Ger-

FIGURE 19. Renaissance jewel; the tiny stars are gold paillons. (*Walters Art Gallery, Baltimore*)

FIGURE 21. Pilgrim bottle, Italian, Venice, late fifteenth–early sixteenth century; enamel on copper. (*The Metropolitan Museum of Art, gift of George Blumenthal, 1941*)

FIGURE 20. Pendant with Diana riding stag, French, sixteenth century; shows encrusté enameling. (*Walters Art Gallery, Baltimore*)

many, and Bohemia. Bohemian enamels proceeded to more and more brilliance, making use of reflected surfaces with translucent contrasting colors or many variations of the same color. These tended to be much too gaudy at times and progressively more extravagant. The Bohemian workshop employed the use of precious stones and bright enamels with an overabundance of filigree, resulting in objects of poor taste. After painted enameling became known, a change in much of this gaudy type of enameling was evidenced throughout Europe.

15. *Swiss*

The Swiss are still producing enamels today, and some of the finest "frits" or unground raw enamels are manufactured in that country. From 1575 to 1650 Geneva was the center of expert research and amazing production of enamels. These enamels were different in character and so sharp and precise that few other countries could compete. The Swiss enamel powders were ground extra fine and applied in the form of paste. The results as seen on watch faces, icons, and minute objets d'art were lacking somewhat in the true translucent or limpid qualities so unique to this medium. Reproductions of portraits and paintings were the cause for esthetic deterioration and little importance can be claimed for mere ornamental jewelry.

16. *English*

In one of the smaller boroughs of London on the south side of the River Thames, a place called Battersea became prominent for its production of enamels. Battersea enamels have become collectors' items, some of which have considerable charm and refinement. These objects, which were evidently produced in large quantities, consisted of enameled snuff boxes, lavalieres, jewel cases, mirror frames, lamp bases, and mementos of every description. Detailed flowers, birds, scrolls, and pastoral scenes often painted in brilliant colors on an opalescent white ground, elegantly finished with gold hinges and edgings characterized these small bibelots in vogue at the time. Battersea and Staffordshire workshops existed in the year 1750 and continued to flourish until 1820. This was a period of superficiality and wanton extravagance, particularly among the wealthy families of nobility. The fine old Byzantine enamels, and the traditions set by the Medieval, Gothic, or Renaissance craftsmen seemed to be entirely forgotten. Beginning about the middle of the nineteenth century, however, a group of artists and artisans who called themselves the "Pre-Raphaelites" tried desperately to revive a more esthetic recollection of previous styles. Their competition with mechanical processes such as machine turning, metal stamping and spinning was never entirely successful. The term "handicraft" and a movement called "arts and crafts" began to be associated with products of questionable taste, and the more bizarre, painted type of English enameling fell into disrepute after the turn of the century.

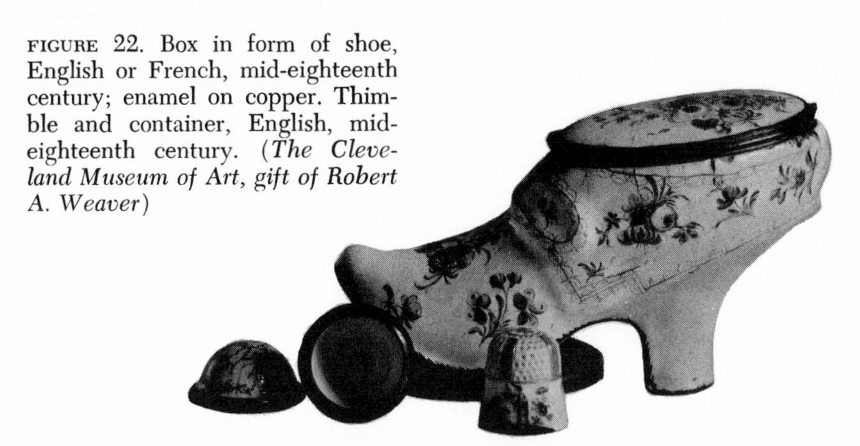

FIGURE 22. Box in form of shoe, English or French, mid-eighteenth century; enamel on copper. Thimble and container, English, mid-eighteenth century. (*The Cleveland Museum of Art, gift of Robert A. Weaver*)

FIGURE 23. Snuff box, French, nineteenth century, by Alexandre Leferre. (*Cleveland Museum of Art*)

17. *Russian, Fabergé*

It would be decidedly erroneous, and certainly unfair not to mention one of the masters of enameling who lived during this epoch. He was Peter Carl Fabergé who was born in St. Petersburg, Russia, in 1846 and died an exile in Lausanne, Switzerland, in 1920. He was indeed one of the finest jewelers and goldsmiths of that time and few craftsmen have been able to compare with this so-called "applied artist." It is claimed that he never actually performed the work himself but delegated the carrying out of his knowledge of the art to an extensive staff of highly-trained workers in his workshops. We can, nevertheless, credit him with an uncanny business sense and the ability to establish the name Fabergé as legend. He understood, mastered, and perfected the most complicated techniques and has been called an expert, superior to Cellini. A pioneer of the art nouveau movement, Fabergé, unlike many of his contemporaries, made extensive study of earlier craftsmen, particularly the work of the old Russian, Greek, and Chinese enamelers.

Naturally not all of his work is appealing today because of the flamboyant style, the superimposed decoration and embellishment; however, one cannot help being impressed or actually awed by the expert workmanship. In his studio he kept a series of 144 tests of transparent enamels, each being shown on engine-turned or hand-engraved metal surfaces. These reflecting surfaces known as "guilloche" patterns became one of the Fabergé trademarks; in fact, his work was always done on an engine-turned base of gold or silver. An exquisite round box done in the French classic manner with gray translucent enamel displayed miniature portraits for his client. Such a remarkable palette of color plus the technique mentioned above gave an iridescent shimmer to all the enameled surfaces of Fabergé works. Snuff boxes, cigarette cases with repoussé, clocks in translucent pink enamel, picture frames, powder boxes, scent bottles, and bonbonnieres (to mention a few of the creations from the studio of this prodigious artist) were constantly in demand. It was an age that could afford him, and he made use of every opportunity. Sometimes his creations consisted of enameled eggs with a surprise for the royal commissioner when the egg was opened. Fantastic creatures similar to those conceived by the surrealist painter, Heironymus Bosch; realistic flowers that appear to be in a glass of water; miniature gold and enameled chairs that have simulated brocaded upholstery and that open up on triple hinges are only a few of the Fabergé pieces in museums or private collections. Fabergé boxes are famous for their hinges. He employed Swedes and

Finns who were supurb craftsmen and often did nothing but fashion hinges. He was fascinated by different colors of gold and experimented with a variety of alloys such as 14 carat gold and pure copper for red gold, or 14 carat gold and pure silver to produce green gold, and also ways to achieve a matt gold finish. Often a piece of work was done in three or four shades of gold.

Much of the Fabergé work was of a technique known as *en plein* enameling, a method of firing a smooth covering over large fields, or sometimes over model surfaces. Even his method of polishing with wooden wheels gave rare and unusual qualities never seen before his time. After the Russian Revolution in 1917, Fabergé's workmen were dispersed. The glittering court and the long tradition were gone. It will be rare indeed if this century produces another enamelist in the class of Fabergé.

18. *Austrian*

The Palais Brenner was the birthplace of the old Vienna School of Arts and Crafts, which for many decades focused its attention on studying the past. Then in 1909 a principal of the school, Alfred Roller, started working for a reformation in the school's attitude. By 1929 when the Jubilee Exhibition of the School was held in the Austrian Museum of Art, it was evident that the meaning of the word *Kunstgewerbe* (art-craft) had changed. To quote Mr. Roller who explained the true relation of handiwork to industry, of mechanical to personal labour, "The change from the drawing-office to the workshop, from knowledge to achievement, from mere exercise of skill to self-expression is the evolution of the Vienna School of Arts and Crafts."[2] Enameling was only one of the crafts influenced by a new kind of thinking, a new kind of acceptance of the mechanical processes—an acceptance that played with, rather than against, the individual craftsman. Students from all over the world studied at the Vienna school and in turn disseminated a new and fresh start in the field of enameling.

19. *American*

In the second quarter of our century, about 1930, enameling on metal suddenly seemed to have emerged in America from what was a fairly obscure medium to one of considerable popularity. Before that time it

[2] L. W. Rochowanski, *Austrian Applied Art* (Wien: Verlag Heinz & Co., 1930).

was rare, if ever, that classes in enameling were taught. There were individual craftsmen who did enameling as an adjunct to jewelry work. There was also the commercial jeweler or watchmaker who occasionally fired enameled objects, but very few who used enameling as a vocation. Today the picture has decidedly changed. Enameling has been accepted as part of numerous high school curriculums, and for many years American art schools, craft schools, and universities have offered courses in enameling. Another phase of recent development, and one less pleasing, is the widespread development of enameling on the amateur or popular level. This phase may have dwindled some in the last decade, and was due, by and large, to socio-economic changes, new uses of leisure time, or a search for self-expression.

Certainly there are many practicing in the field of enamel, those who prefer to call themselves "artist-craftsmen" or "designer-craftsmen," who are cognizant and appreciative of techniques handed down through the centuries, but who, at the same time, are constantly in search of new ways of enameling to express the thinking of the present time.

From a tiny enameled snuff box commissioned by a Duchess of the French court during Louis XIV's reign to one of our modern enameled murals covering many square feet, or from the refined, hand-polished surface of a Fabergé cigarette case to a contemporary rough-textured, free-flowing, burned-edged, amorphous abstraction is a long step, but it is my contention that this fascinating medium remains a challenge to the creative mind and will pass through many phases in the centuries to come.

CHAPTER 2

TEACHING ENAMELING

1. *Preparing the Classroom*

Having accepted the challenge of teaching a course in enameling—whether at the high school, junior college, or university level—the experienced craftsman may feel quite bewildered about establishing such a course if he is doing it for the first time. For the course to be successful the teacher must have a clear picture of the aims of the class, what can be expected in a given length of time, and how to regulate the expenditures allowed within his given budget.

We will make our plans with the idea in mind that the class will be made up of about 15 students. This should be accepted as a full, but not overcrowded, class. Any number of students over 20 becomes a burden to the teacher and disallows time for individual instruction.

Let us begin with the room or studio facilities. Before purchasing equipment, consider first of all the electrical outlets to the room, their location and number. The electrical outlets, wiring and voltage load are of primary importance. Plan for adequate light source in the form of either individual fluorescent or incandescent desk lamps with 100 watt bulbs. Otherwise, ceiling lights of high enough wattage to flood the room completely with light are essential. Even though some of the work will be done by daylight, still plan for evening classes, which have every possibility of gaining in popularity. As the work progresses into a desire on the student's part to do more close and detailed enamels, the individual light source becomes more appreciated. The light should not be placed over the student's right or left shoulder as in reading because naturally the hand casts a shadow on the work. If possible, the light should be directed onto the work from a source in front of the student with, of course, ample shade for the eyes.

There must also be electrical outlets for two or three kilns, a buffing motor, and possibly an electric drill and band saw. Most kilns are manufactured to be used with the regular household current of 110 volts, but if two or more kilns plus buffing motor and other electrical equipment are in the room with the possibility of being used simultaneously it is wise to have the kilns run from a 220 volt line.

It would be unpolitic to specify any particular make of kiln (there are many excellent kilns of all types on the market now) but it is advisable to consider certain sizes and types.

For the class numbering 15 to 20 students there should be at least two kilns and, if possible, one small inexpensive test kiln, although the latter is not absolutely necessary. Never purchase a small kiln with the intention of later investing in a larger one. This is not wise planning. By a "large" kiln we are speaking of one with an interior chamber of at least 12″ × 12″ × 6″. A kiln measuring 12″ × 16″ × 8″ is an even better

FIGURE 24. Enameling classroom plan.

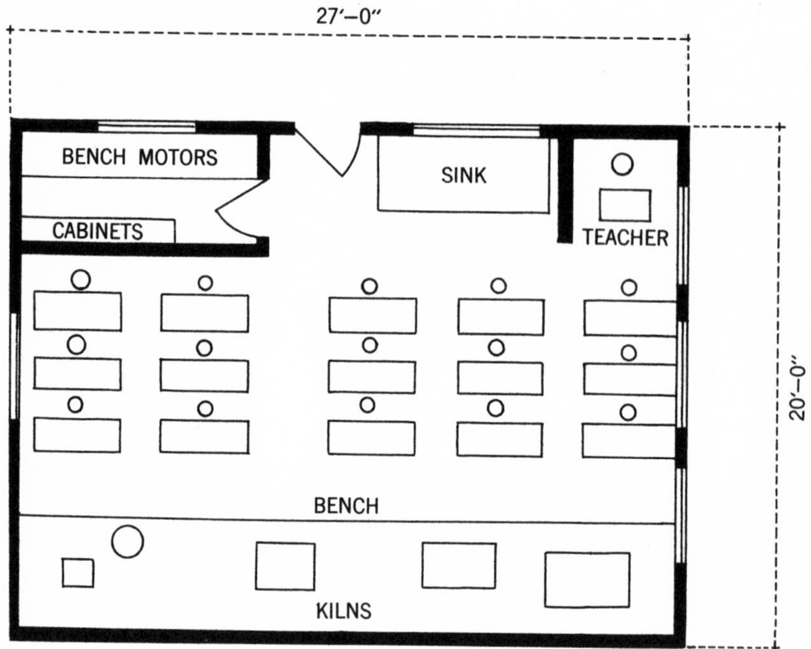

investment. In the large kiln the students can enamel both large and small pieces with ease. The pieces can be fired nearer the back for more immediate and intense heats such as overfiring, and also, pieces placed near the door can be fired at a controlled heat and constantly watched by opening the door slightly from time to time.

As to the type of kiln, for the individual craftsman, this is a matter of personal preference. But for the classroom when 15 students, both beginning and advanced, may keep the kiln in constant use it is of the ultimate concern that the kiln be of the type that guarantees minimum heat loss. Such a kiln would have a door that is best suited for strenuous classroom use. In a large kiln, if the door is of the type that opens and locks from the right-hand side, in order to subject a large 12″ plate or panel to the firing chamber, the door must be completely opened and held open long enough to enter the piece. Balancing the heavy piece on precarious trivets, lifting it with one hand while opening and shutting the door with the other hand lead to considerable heat loss. In the classroom this is ill-considered and disappointing to the next student in turn. If the kiln is of the type that has a door not manipulated by hinges on the left and handle on the right but one that is raised and lowered vertically by a counter-weight the completed act of door-opening is of shorter duration, and the loss of heat from the kiln chamber is subsequently kept at a minimum. This observation is a personal one with me and although it is based upon what I believe is pedagogically sound, I am still loathe to be too dogmatic about it. In many classrooms there have been both types of kilns, and even though the students may have to wait for a kiln to regain its heat, some will still show a preference for a certain kiln. Make sure that the kiln you purchase has a pyrometer and that it is properly calibrated before the students start to work. In the classroom this is of utmost importance.

After the lighting, electrical outlets, and kilns have been decided upon, plan for sinks, polishing motors, and a variety of tools and equipment that are of general use and should be supplied by the school.

First, there is the sink and water supply to be decided upon. It is advisable, although not essential, to have a sink made of some acid-resistant material such as stainless steel, but unless extreme caution is exercised, there is no material that will not eventually be affected by constant exposure to strong nitric acid. Even though the sink is somewhat acid resistant, the plumbing is still in jeopardy as it is customarily made of either iron or copper, neither of which is impervious to acid.

It might be advisable at this time to describe the method of disposing of dirty, overly-used acid. One method is to fill the sink or deep set tub at least three-quarters full of cold water. Then unplug the sink and pour a small amount of used acid into the water to make a very weak dilution while draining. Continue this process until all acid is flushed away. Another method is to turn on both taps (if there are two) and with the faucets fully open carefully dribble the acid down the drain in a tiny stream. After this, continue to flush the water through the plumbing for at least five minutes. *Never,* under any circumstances, pour the acid bath, regardless of how deteriorated it appears, directly down the drain without adequate flushing. Porcelain sinks will be stained by acid and eventually become disintegrated. They must be cleaned with a good cleansing powder after each class session.

Nitric acid, which is purchased in gallon quantities, should be furnished by the school. Needed also are two covered Pyrex dishes approximately 10″ in diameter, and at least 4″ deep, one smaller Pyrex bowl about 7″ in diameter and 4″ deep with a cover, one open low Pyrex baking dish approximately 8″ × 11″ × 2″ deep, sticks and rags for acid swabs, copper or wooden tongs, a box of baking soda in case of burns, a Pyrex measuring cup, and a quart Mason jar for holding the swabs.

When planning the seating arrangement in the enameling studio, remember to allow for an uncluttered passageway from kilns to sink. This will be appreciated when red hot copper has to be taken from the kiln and immediately doused in cold water at the sink for annealing.

In deciding on a buffing motor, remember the problem of the unhealthful and annoying dust from the polishing rouge and the lint from the buffs. One way to solve this problem is to place the buffing equipment in a partitioned-off area or a side room. If this is not feasible, requisition the school to purchase a motor with a built-in dust collector. Spun glass filters, which assure clean operation and collection of dust, protective hood, and lights complete the unit. One buffing motor powered by from 1/6 H.P. to 1/4 H.P. using 115 volts, A.C., 60-cycle current with speeds ranging to 3450 rpm. should be adequate for a class of 15 students. The motor must have right-hand and left-hand tapered spindles.

2. *Preparing the Materials*

It is hoped that the classroom will be set up somewhat along the lines of those suggested, although of course there will have to be many variations and compromises. The next problem is the purchasing of materials to make the projects suggested in our planned course. The following list of materials assumes you will be starting from scratch, which, of course, is seldom the case, as jewelry, metal-working, or other related courses have often preceded a course in enameling. Consequently, there are some kinds of equipment available to the enameling teacher, but, by and large, the class will warrant the purchasing of the following materials:

one or two solid felt buffs 4″ × 1″
several stitched muslin buffs
six Brightboy wheels 3″ × ½″
one rounded felt cone
two cakes of bobbing compound
two cakes of tripoli
one stick of red rouge
six sheets of emery cloth with 2/0 grit
two packages of 00 steel wool
six 8″ flat files with #2 cut
three 7″ long tin snips
a hand drill and assorted set of drills
a jeweler's saw frame with six dozen #0 and six dozen #2 saw blades
a bench vise
a bench pin and anvil combination
four pointed wooden mallets
a planishing hammer with square and round flat heads
two sand bags
three firing forks
one pair of asbestos gloves or mittens
one iron tongs
a variety of enameling stilts
six trivets
two wire mesh firing racks
a porcelain or agate mortar and pestle

Five pounds of 15 to 20 enamels should also be purchased. These might include 10 opaque and 10 transparent colors plus double amounts of flux, opaque black, opaque hard white, and counter enamel.

The teacher can allot the enamels in one of two ways. One, he may place two ounces of each enamel in 3″ × 5″ envelopes and distribute them, or, two, he may simply open five-pound bags of each color, allowing the students to measure their own colors as they purchase the (one level tablespoon is equivalent to one ounce). Sometimes the enamels are sold by the school bookstore. Silver and gold foils as well as metallic lusters are available at most enamel suppliers. Don't forget jars for holding the enamels, enamel tools, small bottles for dusting, wire mesh both #80 and #100, atomizers, rags, and containers for water.

After the problems for the course have been set up, a selection of spun shapes may be purchased from one of the many enamel supply companies throughout the country. Also buy about 15 feet of soft rolled copper 12 inches wide and 18 gauge thick. It is paramount that only the purest copper (99 per cent copper, sometimes called "electrolytic" copper be bought for enameling. A piece of fine silver 18 gauge thick and approximately 4″ × 6″ should be enough to give the beginning class a taste of enameling on silver.

Silver cloisonné wire is perhaps the most popular commodity for the enameling class. Not only because it is less expensive than expected, but because the cloisonné process is one of the most compensating of all techniques. For the class of 15 to 20 students the teacher should order at least 250 feet of fine silver wire .010″ × .040″ or 30 gauge by 18 gauge.

3. *Preparing the Course*

To plan the course, you must first know the amount of time available, which varies according to the student level. The teacher may expect the amount of time to be allotted for the course at given levels to be as shown in Table 1.

The amount of work that can be expected from the average student is a decision the teacher must make. This planned amount of work may have little relation to what he, as a practicing or professional enamelist, is accustomed to produce. However, he must learn neither to underestimate nor to overestimate the student's enthusiasm upon being introduced to a new craft. I can only suggest a sequence of problems showing their proper developmental progress from the known to the unknown. Enameling is not an art to be done rapidly or approached in a desultory manner. It is traditionally a time-consuming craft, and unless the school is a trade school or one that emphasizes job placement, the students need not be coerced into productivity.

It is my firm belief that after a short discussion regarding the historical significance of enamels, followed, perhaps, by showing the students a few examples of good contemporary work, the sooner they can actually begin their work, the more contented they will be.

Below are suggestions for ten problems for beginners, which can be developed, or expanded, to interest the advanced student as well.

TABLE 1

	HIGH SCHOOL	JUNIOR COLLEGE	ART SCHOOL	ART SCHOOL SUMMER SESSION	UNIVERSITY SUMMER SESSION	ONE WEEK WORKSHOP
Laboratory hours per week	7½	3	6	15	10	42
Extra laboratory hours allowed per week	1	4	6	10	24	18
Weeks in school year	36	32	32			
Weeks in summer session				6	6	1
Total hours per week	8½	7	12	25	34	60
Total hours per year	306	224	384			
Total hours per summer session				150	204	
Total hours per one week workshop						60

Problem I OPAQUE AND TRANSPARENT TESTS

1. DESIGN MOTIF: Simple bands of color

2. AIMS:

   a. *To develop design:* This problem does not include design development
   b. *To learn techniques:* Wet inlay application, firing enamels over metal foils
   c. *To develop skills:* Handling small objects, control of tools, placing enamels in the kiln, learning effect of various heats

3. LESSON DEVELOPMENT:

Let us start by making tests of all the enamels in our palette, both transparent and opaque. The teacher must be adamant about testing enamels, finding a way to convince the students of its importance, and holding their interest until the job is finished. The making of tests will continue as long as the enamelist carries on his work. Small transparent-domed kilns are convenient for this work, but for special-color tests a calibrated kiln is essential in order to know the exact temperatures used. Consequently without the pyrometer the small kiln is of little avail. This first problem of testing each color makes our subsequent work in enameling more significant and keeps it from becoming mere hobby work.

The opaque tests (Figure 24A) need space for only two variations of the color. Bend a one-inch square tab of copper in a shallow arc, file its edges, drill a hole at the top for hanging, and prepare for enameling by immersing in a solution of one-part nitric acid to four-parts water. *Always add acid to water.* Follow by rubbing with 00 steel wool and saliva or by rinsing quickly with a liquid detergent. Apply the counter-enamel and fire. The arc-shaped tab will serve two purposes: one, the enamel never touches the kiln except at the very edge of the counter-enamel; and, two, it is easily lifted with a small spatula or ordinary kitchen knife.

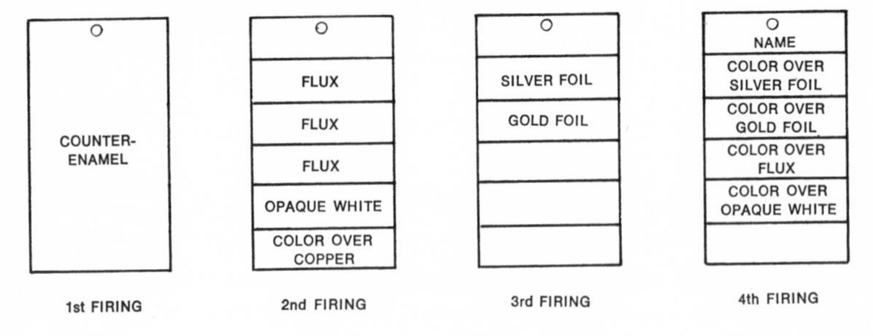

TEST FOR OPAQUE ENAMELS

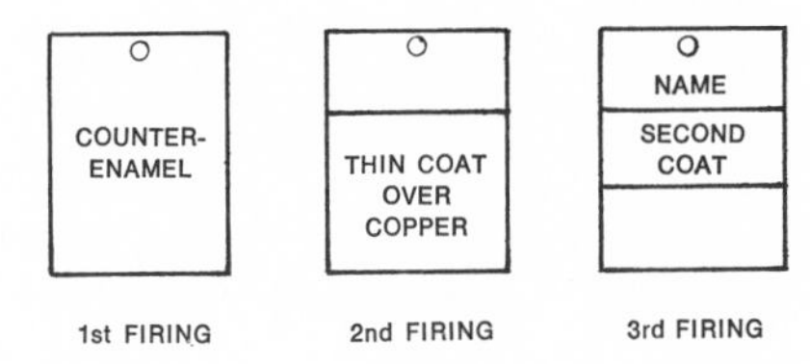

FIGURE 24A. Diagram showing steps in firing the transparent and opaque enamel tests.

The convex side of the test is now cleaned and prepared for enameling. Place the opaque enamel over the entire front surface. For experimentation, give the opaque an overfiring (about 1650° F.).

In overfiring the opaque enamel on your test you will find that often there will be darkened areas, as in the case of red and vermilion. Other colors will become distinctly transparent. Leave one-half of the tab in the overfired condition for your palette of effects and apply another, slightly thicker, coat of opaque to the remaining half. Fire at a lower temperature than before (about 1450° F.) and you will have as a result a complete test of one opaque color, showing it in both its true and its overfired state.

Both silver and gold foil can be bought in booklets containing 12 or more sheets, each approximately 4 inches square. (The areas of silver foil used in enameling are called "paillons.") Prepare a piece of copper ¾ inch wide by 1½ inches long for the transparent tests. Any one of the transparents is first fired over this copper tab in a smooth coat. Now the shape of the paillon is traced onto a piece of tracing paper and the silver foil is held between it and another piece of tracing paper, then cut with sharp shears or a razor blade. The paillon is next punched with many pinpoints, preferably 150 to 200 per square inch, so as to allow air to escape from under the foil when it is fused to the enamel.

Adhere the paillon to the enameled surface with gum tragacanth solution and carefully absorb any excess moisture with a blotter or soft rag. When the paillon is completely adhered to the enamel, it is fired sufficiently. Take it out of the kiln during the process of firing and burnish it with the back of a dull kitchen knife or artist's spatula. Caution should be taken not to overfire the paillon as it is quite possible to burn it, in which case it is ruined and will have to be stoned off with a carborundum stick.

Making transparent tests on copper involves four separate firings (Figure 24A).

Firing No. 1 is the counter-enamel firing. Use scale-off to protect the front surface. Give just enough firing to set the counter-enamel as it will be fused further in the subsequent firings. After cleaning the front surface, prepare to show the transparent color as it looks over silver foil, gold foil, flux, opaque white, and copper. This will involve three separate firings.

Firing No. 2 has the color applied directly over copper for the area at the bottom. At the same time an area of opaque white and three areas of flux are placed above. Fire, clean the edges with file or emery cloth, and wipe off the surface in preparation for the next step.

Firing No. 3 enables you to place the silver paillon and the gold paillon over two of the flux areas in the manner described above. Fire as before, being sure that the paillons are adhered perfectly to the enamel.

Clean and proceed to Firing No. 4, which is the last firing, giving us the color effect over opaque white, flux, gold, and silver.

PROBLEM II A SIMPLE SMALL TRAY WITH WET INLAY AND FOILS

1. DESIGN MOTIF: Animals, seed pods, flowers
2. AIMS:
   a. *To develop design:* Designing within a circle, balancing colors, rhythms and shapes
   b. *To learn techniques:* Application of wet-inlay method and processes learned in Problem I
   c. *To develop skills:* Control of enamel on rounded form, firing with trivets

FIGURE 25. Problem II, a simple small tray with wet inlay and foils; motifs: animals, seed pods, flowers.

3. LESSON DEVELOPMENT:

Having learned the spatula or wet-inlay process and also the dusting and metal-foil procedure while making the tests, the student is now ready to try his hand at a simply designed small tray. Some subject should be suggested for a motif, such as abstract animals, leaves, seed pods, or flowers. Demonstrations of rhythm, placement, and relationship might be given at the blackboard by the teacher. (Technique is not to be a substitute for poor design, and the teacher must do everything in his power to prove this point. He must aid, suggest, inspire, encourage, motivate, and stimulate but never dictate or dominate the student's thinking. He should accept accidents and naive technical errors, which have a way of becoming "important discoveries" at this stage of the game.) The student will enjoy an initial thrill as he completes his first enamel piece—finishing and polishing the edges by carefully filing to remove enamel and fire glaze, using the fine or medium file, then buffing out all file scratches with the felt buff and tripoli, and after removing the tripoli with a cloth or ammonia, finally polishing with the muslin buff and jeweler's red rouge. The wise teacher will utilize the student's emotional reaction to his first piece and build constructively from this point on.

PROBLEM III LARGE FLAT PLATE WITH SGRAFFITO AND STENCIL

1. DESIGN MOTIF: Geometric shapes and lines

2. AIMS:
   a. *To develop design:* Study of direction, balance and proportion of geometric relationships
   b. *To learn techniques:* Study various uses of sgraffito and stenciling
   c. *To develop skills:* Calligraphic ability, stencil efficiency

3. LESSON DEVELOPMENT:

Now that the first enamel piece has been accomplished and many of the initial difficulties have been overcome, it is time to progress to other techniques and more serious problems of design. As each problem is presented, the student should gain in self-confidence and should try to develop his own design conceptions in a way that is his and his alone. The enameling studio is not a place for communal chatter, but a place for individual thinking. Friends should not open kiln doors for one another, nor aid in designing one another's problems. However, the benefits of an interest in common and of rapport between teacher and student can be gained by the skillful manipulation of discussion at the time of general criticism. A good plan might be as follows: Preceding each new problem the students should present their sketches and ideas in full color, using only hues and values that are obtainable by directly referring to their tests. These in turn may be placed on the board in front of the room for a serious discussion by the instructor in charge. Students are apt to make greater progress from these community discussions of their own work and the work of their fellowmen than they are from hours of cut-and-dried lectures by the instructor.

The third problem is a larger flat plate of approximately 7″ or 8″ in diameter. This design will involve experimentation with many types of sgraffito. The motif is geometric in style and for this problem we will exclude the gold or silver foil technique. Sgraffito from dark through to light, from light to dark, from opaque to transparent, or transparent sgraffitoed through to opaque are but a few possibilities. Combine these effects with flat areas making use of the contrasting stencil technique. These stenciled supplementary colors are to be added after the sgraffito is accomplished and fired, and always with transparent enamels. Use moistened paper toweling for stenciling. Keep a very

FIGURE 26. Problem III, large flat plate with sgraffito and stencils; motif: geometric shapes and lines.

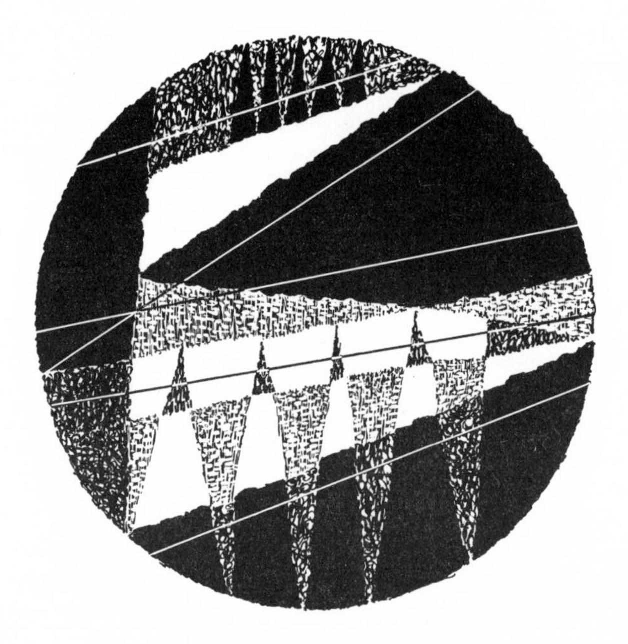

thin—in fact, most delicate—dusting with several built-up tones and subsequent firings for the best result. To avoid a hard-edged look to the stenciled areas try tearing instead of cutting the paper or lifting a cardboard stencil an inch away from the enameled surface. Avoid an effect of pile-up or thickness at the edges of these areas. It is always advisable to think of enamels as "thin skins" of glass fused to the surface of metals rather than thick lumpy surfaces similar to the frosting of a cake. Finish the plate in the same manner as the tray in Problem II.

PROBLEM IV DEEP BOWL WITH FIREGLAZE AND LIQUID FLUX TEXTURES

1. DESIGN MOTIF: Abstract or nonobjective approach

2. AIMS:
   a. *To develop design:* Serious consideration of abstract and non-objective textural surfaces

FIGURE 27. Problem IV, deep bowl with liquid flux textures, by Eleanore Slobin. (*The Cleveland Museum of Art, lent by the artist for the forty-eighth May Show*)

b. *To learn techniques:* Study possibilities of developing textures by fireglaze and liquid flux
c. *To develop skills:* How to apply a flowing medium and control of fireglaze.

3. LESSON DEVELOPMENT:

A deep bowl with more or less vertical sides brings up new technical problems. Dusting one section of the bowl at a time while holding it in a slanted position is a profitable method to use. Adhere the enamel carefully using a more concentrated solution of Klyr Fyr or gum tragacanth, being careful not to drip or run at the edges of the bowl. This technique must be mastered at an early stage.[1]

The design for this problem is to be completely nonobjective in character, that is, no approximation or symbolism of recognizable objects is to be incorporated. There are two methods that will produce

[1] Polly Rothenberg, "Enameling Steep-sided Bowls," *Ceramic Monthly*, June 1965, pp. 15–17.

FIGURE 28. "Tulip and Bud," plate by the author showing liquid flux technique.

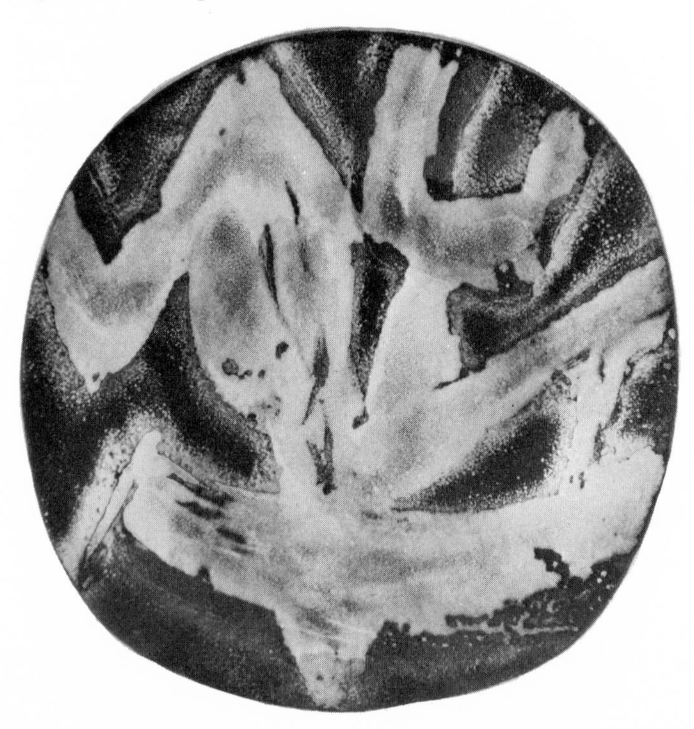

textural effects. One uses what is known as liquid flux.[2] This is a finely-ground flux suspended in a solution that is volatile and that allows the flux to adhere to the surface of the metal. This medium may be painted, brushed, spattered, sprayed or poured onto the copper. By allowing certain areas of the copper to be exposed, and by allowing those areas to become fireglazed when fusing the flux, many free and amorphous shapes can be achieved. Interesting flowing patterns of endless variety become most fascinating. Follow this firing (or firings) with subsequent thin coats of transparent colors over the entire surface.

A second method of obtaining a textural composition is by purposely burning the copper to obtain a fireglaze or oxidization surface. This surface has a tendency to chip off in parts, often leaving a fascinating textural pattern. Such a technique is not new and no doubt occurs the first time the student subjects his exposed copper to the heat of the kiln. It is, however, a matter of chance and a matter of choice on the part of the student in selecting a plausible design. Try firing the copper at about 1550° F. for one minute for such an effect.

The problem involves combining these two techniques, one which can be controlled and one which is purely arbitrary, with the intention of producing a pleasing arrangement. The freedom allowable in this problem is directly planned in the curriculum to relieve the student of the careful planning and preconceived sketching that accompanied the previous problems. However, the teacher must guide the student to some extent, at least to a point where he recognizes fitness and compatibility of pattern to form as against meaningless or unattractive doodling.

PROBLEM V HAMMERED FREE-FORM WITH SURFACE DECORATION

1. DESIGN MOTIF: Abstract fruits, flowers, birds

2. AIMS:
   a. *To develop design:* Study how to treat subject matter abstractly, how to relate decoration to a unique shape
   b. *To learn techniques*: New possibilities of "dump-off" and overglaze
   c. *To develop skills:* Principles of simple hammering, creating an original form, control of overglaze lines

3. LESSON DEVELOPMENT:

The next problem is a hammered free form. The technique is not

[2] Helen Worrall, "Enameling with Liquid Flux," *Ceramic Monthly,* May 1965, pp. 18–21.

difficult, and the instructions for hammering should be kept very simple and direct. Although a copper form for enameling requires less attention to surface refinement than does a beautifully plotted and carefully executed silver bowl, this does not mean that the copper form can be too imperfect or too crudely conceived. However, the final result can be achieved with a few simple tools, namely, the pointed or round-ended wooden mallet, the sand bag, and a few files. Start the demonstration by showing the students how to sink the base area, and then by hammering on lines leading directly to the points or corners of the tray, slowly bring it into its desired form. Because a clear concept of the shape before hammering will enable the student to strike more meaningfully, rather than pounding away indiscriminately, many sketches should be made previous to the cutting of the shape. These sketches should be subject to class discussion, the teacher pointing out the difference between subtle, simple shapes of good proportion and shapes that reflect the trite, commercial appearance.

Refinements of the form follow after annealing several times. Think in terms of sculpture or modeled copper that can be cupped, domed, or stretched rather than merely bent in the manner that paper bends. The subject for design motifs is fruit, flowers, and birds. It is doubtful that a naturalistic interpretation would be appropriate for an asymmetric free shape, so keep in mind that the subjects for design must be completely abstract in character.

For decoration use an enameling technique known as "dumping-off," combining this technique with overglaze delineation. The term "dump-off" means any effect that can be gained by first painting, sponging, or swabbing the copper with some solution such as Klyr Fyr, thick gum tragacanth, or Klyr Cote (these are all obtainable commercially) and

FIGURE 29. A hand hammered free form "dump off" and overglaze decoration.

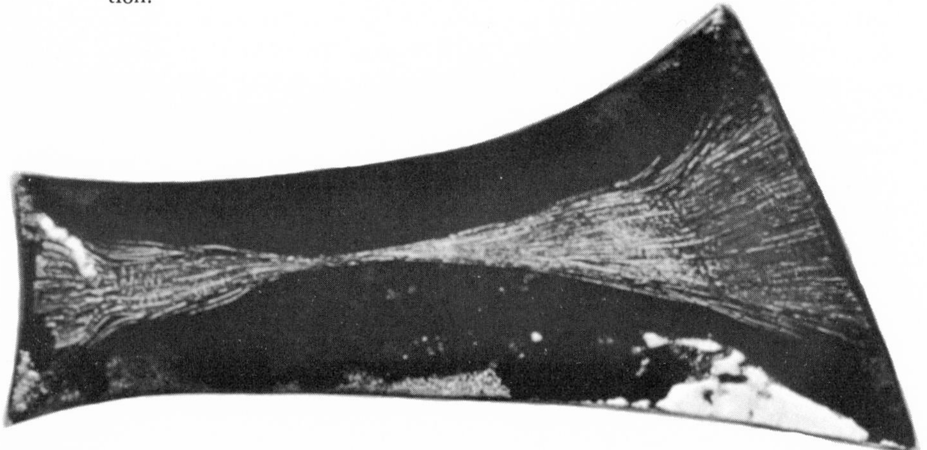

then proceeding to dust on dry enamel over the partially wetted surface. By dumping off any excess enamel that does not adhere, purely organic textures or shapes can be found. Several trials may be necessary before the results are satisfactory. Several firings of transparent enamel applied in the dump-off manner may be superimposed and fired separately enabling the student to arrive at an interesting result.

Develop the design further by working into it with a good black overglaze for delineation and details. Sgraffito through the overglaze to incorporate it into the underpattern. Be sure to regrind the powdered overglaze on a glass or porcelain tile with the spatula using a little oil of sassafras and perhaps a drop of squeegee oil for easier manipulation.

PROBLEM VI SMALL ROUND COVERED BOX WITH SILVER WIRE CLOISONNÉ

1. DESIGN MOTIF: Conventional pattern or ecclesiastical motif

2. AIMS:
   a. *To develop design:* How to use the conventional or formal design in a contemporary manner. Making figures and symbols fit a given space
   b. *To learn techniques:* Bending cloisonné wires to conform to a specific type of design
   c. *To develop skills:* Working in small areas with pointed tools and finely-ground enamels. Control of heats, skill in stoning and polishing

3. LESSON DEVELOPMENT:

The problem is to design and render a small circular box with a separate unhinged cover. (These shapes can be purchased from most enamel suppliers.) Although the class has moved from simple tests to more involved techniques in just a few weeks, probably the cloisonné manner of working is not too difficult to introduce at this time. Silver wire cloisonné is one of the oldest techniques in the history of enameling.

Clean the metal as for tests (Problem I) and bend the wires to correspond to the previously prepared design. Carefully consider both the height and thickness of the wire. I recommend that the wire be 30-gauge wide and 18-gauge high, or .010″ by .040″.

When the first firing has been completed, you will notice that the enamels have shrunk so that they seem to be clinging to the wire cloisons. To build the enamel up to the level of the wire, the plate or panel will need complete re-enameling. Possibly a third coat of enamel will be necessary.

Stone the enamel down to the level of the wires. With the enamel firmly gripped in one hand and a medium coarse carborundum stone in the other, place both hands in a deep bowl of water or under running water, and grind with a circular motion. (Running water flushes away all of these finely ground particles and is by far the better method.) You must stone until all wires are evenly exposed and until all mounds and depressions in the surface of the enamels have disappeared. Occasional drying will reveal to you how much more stoning there is to be done.

Fire now before applying more enamel so that scratches left by the stoning will not show under the last enamel coating. The stoning will

FIGURE 30. Problem VI, small round covered box with silver wire cloisonné; motif: conventional pattern or ecclesiastical motif.

usually reveal some pits and air bubbles. These should be refilled before this firing. Buff the wires with tripoli, using a felt buffer. No damage seems to be done to the enamel by this method.

The teacher should insist upon a very carefully planned full-color sketch with particular attention paid to drawing and refinement of lines. Also an inked line rendering should be made on a separate plate to be used as the student's working drawing when bending the wires. As a relief from the pure imagination required by the previous problem, a great deal of time and serious thought should be spent on the designing before presentation for general class criticism. Styles,

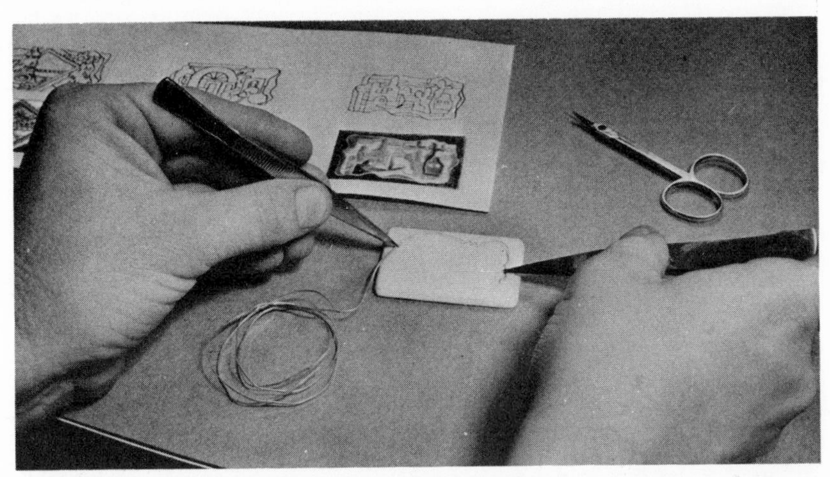

FIGURE 30A. Small tweezers and manicure scissors are convenient tools for forming the cloisonné wire.

FIGURE 30B. Finely ground enamel is carefully placed between the cloisonné wire.

negative and positive areas, color relationships, rhythm, focal interests, and many other basic criteria should be included in the instructor's discussion.

If the design is not ecclesiastical in concept, a series of simple, conventional shapes may be used. In any case, because the cover is not hinged and must be viewed from any position, it is logical to think of the design as formal and based on a central point rather than as asymmetrical. (This approach may appear dogmatic, but by so doing the teacher is introducing one more specific way of thinking.) Conventional design is not less difficult because of its regular repetition or duplication; in fact, it becomes a challenge to create a good conventional design that will have a contemporary look about it.

There are many new technical problems in enameling a covered round box. A safe procedure is to enamel the inside of the cover and box first, making sure that it will still fit, and then proceed to enamel the outside of box, and lastly the cloisonné cover. (For more advanced cloisonné study, see Chapter V, page 130.)

FIGURE 30C. Tiny bits of metal foil, or paillons, are placed in certain areas, later to receive coats of transparent enamel.

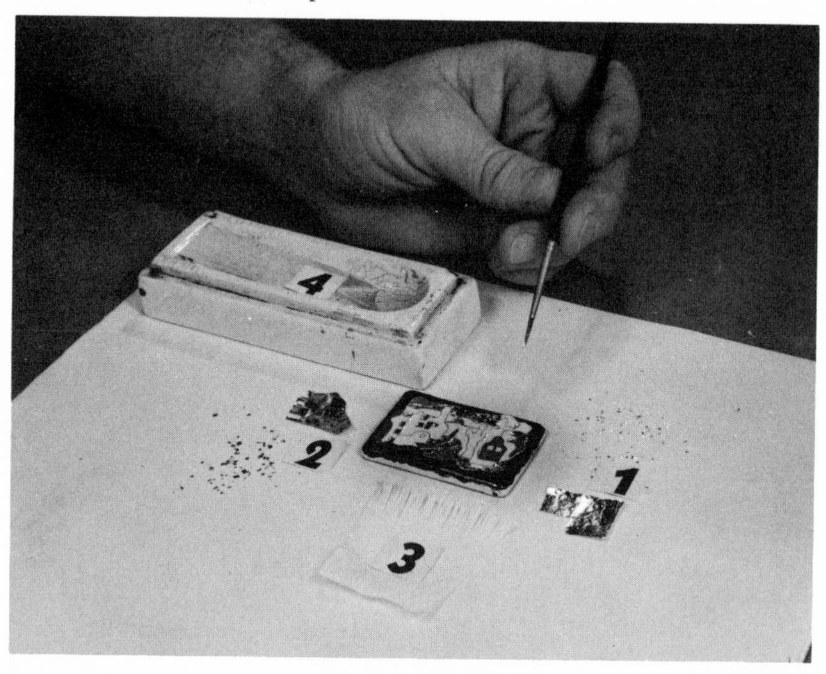

PROBLEM VII ROUND PLATE OR DEEP BOWL WITH SEPARATION ENAMEL

1. DESIGN MOTIF: Spots, lines, masses

2. AIMS:

   a. *To develop design:* Study the placing of simple spots, lines, and masses of color
   b. *To learn techniques:* Research and practice with a flowing or moving pattern
   c. *To develop skills:* Control in firing at temperatures above 1500° F.

3. LESSON DEVELOPMENT:

In contrast to the last problem, which involved many hours of detailed designing, planning, and wire-bending, the use of separation enamel requires much less concentration and is certainly much less time consuming. The use of separation enamel is not in the same category as cloisonné enamel. Its effect is quite accidental and it is the joy of the hobbyist who is weak in design or who has had no formal training in it. However, we should be aware of more recent developments in enameling, without eschewing the older techniques. (If we feel that some of

FIGURE 31. Separation enamel can cause interesting textural effects on a deep bowl.

the quicker and easier techniques have been misused by the dilettante, perhaps we should experiment in the classroom with these and try to raise them to a higher level.) It is just as legitimate for enamel to flow or move on the surface of metal when it is in a molten stage as it is to remain exactly where it was placed. All such expedients as chunks, threads, and separation enamel have their place but, unfortunately, because of their seeming ease of fabrication and the speed with which they can be produced, shops and sales counters are oversupplied with mediocre examples of this technique.

By first firing a good coat of flux or a very soft opaque white on the copper, then areas or lines of light transparent enamels, followed by an entire coat of some darker transparent color before the separation enamel is applied, better results may be achieved. The separation enamel is painted on with a brush in lines or spots, which may either relate or contrast to the underpattern. Finally, the bowl is subjected to a heat of 1600° F., causing the upper coats of enamel to separate and to expose the under coat of flux wherever the separation enamel was applied. This problem will be successful only if the student is willing to make several trial runs and develop effects that could be achieved in no other way. A revision of the original rough sketch involving a simple placement of lines, spots, and masses should be encouraged.

Problem VIII FLAT PANEL WITH COPPER CHAMPLEVÉ

1. DESIGN MOTIF: Figures and trees
2. AIMS:
    a. *To develop design:* Practice in good drawing and use of figures in given areas
    b. *To learn techniques:* Discovering the method of champlevé, types of acid etching, and application of enamels
    c. *To develop skills:* Controling the painting of a resist, and skill with the acid bath and stoning
3. LESSON DEVELOPMENT:

In a beginning course it is of interest to students to try their hand at as many of the basic techniques as possible. Although there will not be enough time to go too far with any one technique, in subsequent years the student who has been introduced to several manners of working may carry on with any of them. This problem involves a small panel in the champlevé technique, the third of the older forms of enameling. The first tray could be called "painted enameling" or a form of the

Limoges technique; Problem VI was cloisonné; and this problem uses one of the most ancient devices, that of etching or gouging out certain areas that are then filled with enamel and fired to the level of the metal.

A piece of copper 4″ × 7″ is adequate for the experiment. Here, again, the color scheme should be planned carefully as the warm copper color must play in juxtaposition with the tones of the enamel. In this panel, the student will be restricted in subject matter to one or more figures with some tree form in relationship. Because the charm of champlevé enamels rests on a rather bold flat appearance, no attempt at realism should be considered. Stylistic contours and proportions would be more appropriate. Allow the asphaltum varnish, which is painted in the negative areas, to dry overnight, and then subject the piece to a mild nitric acid bath with proportions of one part acid to seven parts water. Eight to ten hours may be needed to etch 1⁄32″ deep. In this way the acid "bite" is cleaner and the asphaltum is less apt to disintegrate in the process. Stoning to bring the enamel to the

FIGURE 32. Problem VIII, flat panel with champlevé; (A) Preliminary sketch, (B) Development of asphaltum applied to copper plate. (Note: black represents asphaltum; white, copper.)

level of the metal must be done with considerable care so that the edges of the enameled areas are nicely articulated. After the final firing and a gentle cleaning of the fireglaze with weak acid, it is possible to buff the entire panel using the felt buff with tripoli.

Problem IX A COMPOSITE VERTICAL PANEL WITH FITTED SECTIONS MOUNTED ON WOOD

1. DESIGN MOTIF: Fish, fish nets, and related marine subjects

2. AIMS:

   a. *To develop design:* Practice and study of principles of composition, balance of color and shapes

   b. *To learn techniques:* How to subdivide areas for sawed sections, techniques for mounting and finishing woods

   c. *To develop skills:* Skill in sawing, wet inlaying, foil enameling, and handling details

3. LESSON DEVELOPMENT:

For this major problem choose a size of approximately 6″ × 13″. Start thinking about the subject matter. Such things as fish, fish nets and boats can be very trite indeed, if the students resort to a realistic approach. An enameled panel is not a picture. There need be no semblance of existing horizon lines or obvious perspective. Also, because fish are of themselves stylistic in shape and gay in color, the

FIGURE 33. Problem IX, detail showing fitted sections.

FIGURE 34. "Trawlers with Catch" by the author is composed of 28 fitted segments.

student should search for ways of exaggerating, distorting, or abstracting them for more interest. Exploration and imagination must be exercised in order not to visualize the fish form as a cutout piece of silver foil with literal coloration as found in nature.

Subdivisions of the panel should be such that when the pieces are fitted together these joinings do not disturb the unity of the design. Try to find anatomical or structural lines upon which to work when making these divisions. As much as possible make the shapes simple and bold rather than delicate and pointed (this will obviate any expansion changes as the parts may need to be subjected to the kiln several times). The panel should be mounted first on thin wood about ¼″ high and then on a larger panel with contrasting value by the application of a mixture of stains and finishes. Keep each segment of the panel flat by the use of weights. Guard against piling the enamel high around the edges of the sections. Assiduously stone all parts to bring the composite panel into one final flat plane.

Problem X PERSONAL JEWELRY—TIE CLIP, CUFF LINKS, EARRINGS, KEY RING, BRACELET, EMBLEM, OR PIN, WITH ENAMELING ON FINE SILVER

1. DESIGN MOTIF: Spot, leaf, emblem, monogram, etc.

2. AIMS:

   a. *To develop design:* Taste in choice of simple motif for applied design; appropriate design for function
   b. *To learn techniques:* Enameling on domed shapes; enameling on fine silver
   c. *To develop skills:* Buffing jewelry, skill in simple soldering

3. LESSON DEVELOPMENT:

In the last problem it might be provident to be lenient in the demand for a serious study of enameling techniques. However, in the making of a simple piece of jewelry beware of the gift-shop type of enameled knickknacks. There are on the market a variety of jewelry findings, such as cuff links, pin settings, earrings, etc., for the addition of small enameled areas. If it is possible to discover a few of these settings that are simple and in good proportion, purchase them for this last problem. They might present quite a challenge to the student, allowing him to raise the level of cheap hobby jewelry to something of more refinement. One way to do this would be to form the enameled setting of 18 or 20 gauge fine silver in a dapping block. These, when polished and skillfully enameled with one brilliant transparent color, could become quite attractive. Also, the introduction of lead soldering, or even simple silver hard soldering for bezels, would not be too much to be expected. (An inexpensive Burnzomatic torch would be adequate for the soldering job.) The tiny enameled domes may be attached with epoxy cement when there is no stress, provided a good fitting is planned. The average high school student is inclined to think of personal jewelry in terms of monograms, or class emblems. I see no reason why such an attitude should be frowned upon. It is better to do some research to find dignified and elegant examples in the best craft publications to show the student what might be done than to disallow such a problem as part of the course.

FIGURE 35. Problem X, personal jewelry—tie clips, pins, earrings, and so forth.

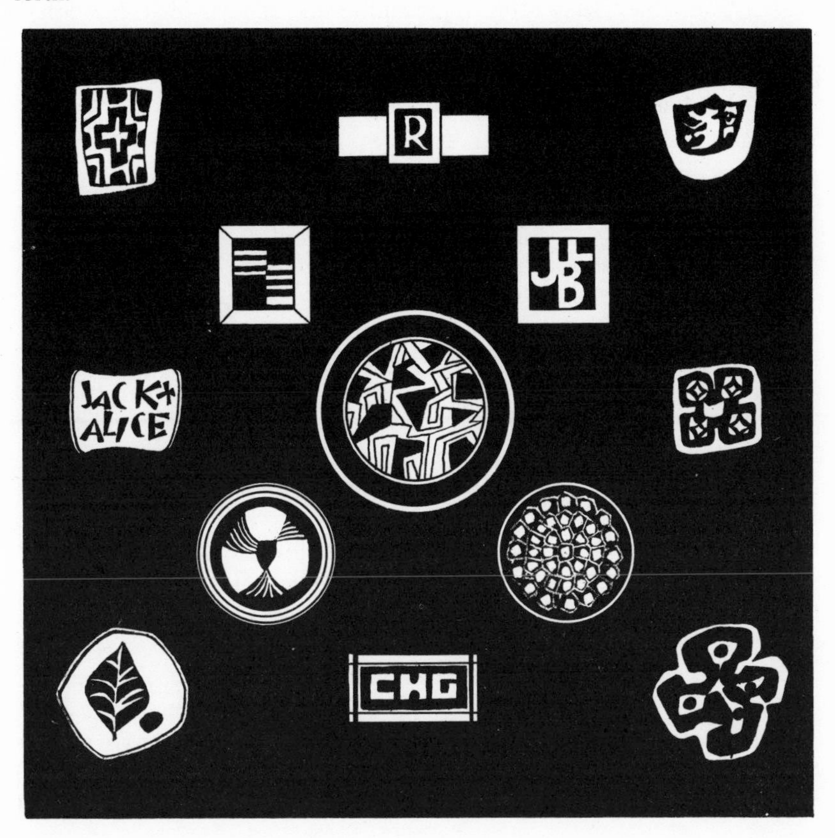

4. *Preparing the Teacher*

How does the teacher of a course in enameling differ from a teacher of sculpture or pottery? Perhaps, for one thing, he must convince the administration that enameling has enough value, both as a practicing craft and as an esthetic experience, to warrant its being included in the curriculum. As I have stated in my chapter on the history of enameling, it was not until after the first quarter of the twentieth century that enameling was taught as a regular course in art schools or high schools. By way of contrast, pottery and sculpture have been taught since art schools were conceived; therefore the young potter or sculptor who is beginning his career as a teacher is not confronted with the same problem as is the enameler. The craftsman who intends to teach enameling should have several convictions. First, he must believe in the historical value of the medium. He must be aware of the fact that the great enamels of the past were something more than man's passing fancy, that they often represented an art form that embodied man's supreme esthetic endeavor. In the Dark Ages before the year 1000, when many of the most magnificent Byzantine enamels were made, the craft acquired an importance of highest rank. During the Middle Ages enameling, like stained glass windows and sculpture, was used as a means of communication with the illiterate peasants. The great spiritual messages of Christianity were repeatedly taught through the medium of enameling in reliquaries and panels. Second, the teacher of enameling must be aware that his craft is one that incorporates the knowledge of color, design, craftsmanship, and to some extent engineering and chemistry.

Teaching a course in enameling is not merely a question of keeping students busy. So often in craft classes, the students prefer to "do" rather than to "think." The teacher must have enough pedagogical skill to utilize the student's urge to work, but, at the same time, promote a logical sequence of study from simple to more complex design problems.

Enameling can be very frustrating. Difficulty with application of enamel, firing time, and controls can cause much disappointment. The teacher has to be patient, courageous, and resolute when teaching students in such an experimental and somewhat unpredictable medium.

The teacher's preparation should be something more than a few easy courses in enameling. He would do well to avail himself of four or five years of art school training. To major in design, sculpture, or painting would be of equal value and equally applicable to the position he hopes to fill. Concentrated study and practice in drawing and color theory cannot be by-passed as preparation for the enameling teacher. Actual knowledge of how to make enamels and endless exploration with the medium appear to be of less importance as a part of the teacher's background than the fact that he is a well-trained artist, a sensitive designer, a demanding craftsman.

The students will gain much from the teacher's philosophy about art in general, about his reactions to current thinking, about his appreciation and evaluation of craftsmanship. It therefore behooves him to approach the study of enameling with the broadest and most encompassing viewpoint. This craft has not reached its zenith; it is still in a probative stage in this country. It remains for the coming generations to develop new uses and to cultivate further advancements with the medium. With such a philosophy about his course, the teacher will promote enthusiasm that inevitably leads to better-than-expected results from his students.

5. *Preparing the Student*

Perhaps as in no other course being taught today, the student's lack of design background is the bane of most teachers. Design background does not mean that a student has been matriculating in this or that course for years only to have accumulated a series of devices and mannerisms that he thinks he will use in enameling. His understanding of design may have come from any one of a number of subjects, such as sculpture, painting, life drawing, or construction drawing. A course in basic design theory should give the student at least a working knowledge of designing for enamels and a means of communication with the instructor. But still, continued formal training is no guaranteed panacea for the poor design student who has no initiative to do research for himself. The student who approaches enameling with scrapbooks full of anecdotal sketches, and ideas from nature and life about him, is the student who will show genuine design promise. Unfortunately, enameling attracts those who wish to take summer courses or workshops for a kind of relaxation, or to pick up an extra credit because the work looks easy. Enameling, as any other form of creative art, is never easy. It is frustrating, time consuming, and enervating. Students must understand this and approach the work with a healthy attitude.

In preparing the student for the work ahead the teacher should develop some kind of philosophy at the beginning of the course. The initial attitude of the student should be established by the teacher. Have each student come to class prepared to work at a clean bench with tools and paraphernalia laid out in order. In fact each student should have a few pieces of white typewriting paper spread out where he is to work with brushes, enameling tools, water, and rag easily accessible. (As pedantic as this may sound, it is certainly not a bad idea, especially as the laboratory time has a habit of vanishing all too quickly. It is simple logic that if 50 per cent of the time is utilized in searching for one's equipment only half as much work can be accomplished.)

The student should be inspired to make full color renderings of each project. Revisions of color, values, drawing, and proportions can be more providently worked out on paper than with expensive enamels and metal. For experimental work such sketching might not be expedient, and, to be sure, many craftsmen prefer the trial-and-error method to making preliminary sketches. For the weaker design students much more can be accomplished when there is at least a rough color drawing to criticize.

FIGURE 36. The proper placement of tools, rag, water, and so forth. Note the use of white paper as a working surface. Folded paper is convenient for pouring dry enamel.

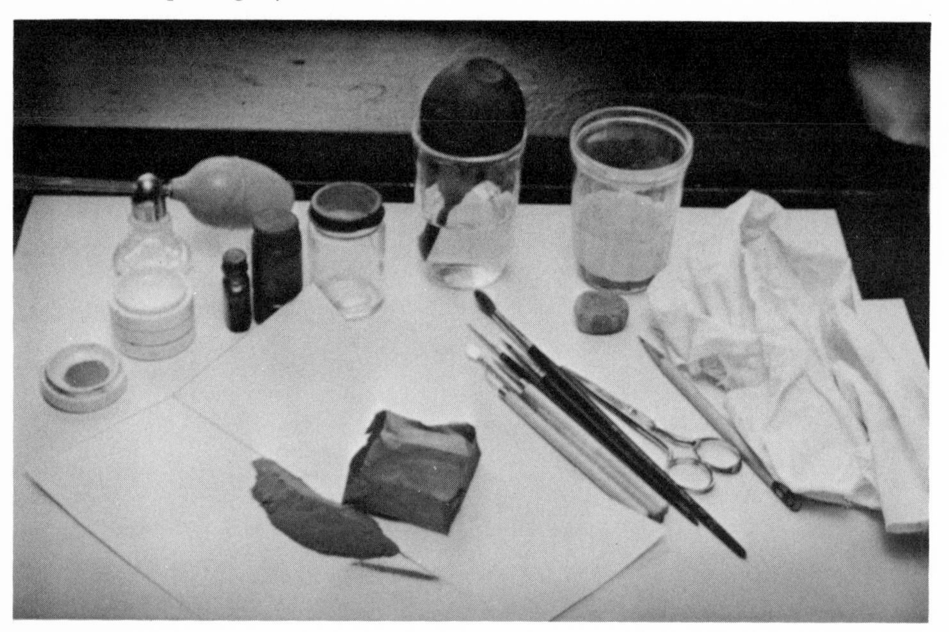

6. *Preparing the Results*

The course is concluded. The students have enjoyed their initiation to a fascinating craft. It is to be hoped that the instructor has enjoyed the experience also and that the results are worthy of his time and effort. Some proof of these results in the form of an exhibition will inevitably be expected of him. The problems of show cases and display are in some ways unique (lighting and background) as regards enameling, but not too unlike the requisites for silverwork or pottery. From the start let the teacher in selecting the pieces he wishes to exhibit be devoid of sentiment and use his prerogative to choose only that which interests him. He has two issues to keep in mind—one, the standard of work he wishes to promulgate and, two, the design of the cases where the work will be displayed. The appearance of the display case reflects upon the instructor's sense of design and taste as much as his teaching does. He must be discriminating in choosing the right number of pieces, not allowing quantity to supersede quality. Fewer examples placed strategically and grouped in an unusual manner are always more effective than a crowded showcase of pieces of questionable merit. Start with the back wall or most distant plane. By hanging those enamels such as panels, pictures, wall crosses, and a few flat plates first,

FIGURE 37. In preparing an exhibit, it is necessary to consider the relation of color, shape, and size.

you are able to visualize the general concept of the composition of the case. Work toward the front of the case, placing the smaller and more important enamels nearer the viewer. The relation of color, shape, and size must be considered as well as negative space, which is often the most important consideration. Enamels are intimate, having an opulence and luxuriousness about them that must be recognized, but at the same time they should never be displayed with ostentation. Elaborate backgrounds and fancifully draped textiles are bound to be in bad taste. For the case, search for a color that will relate to the dominant color of copper and still remain in the background. Enamels must be lighted from the front if at all possible, perhaps by the use of outside spot lights, but glare must be avoided at all costs.

The proper price to charge for student's work is one of the teacher's unavoidable problems. With the changing economy it would be ridiculous to state actual figures here. But, to generalize, let us say that if a retail store sells an enameled ash tray for $8.50 (the artist would probably receive one-half of that amount), then the student selling directly is also justified in charging $8.50. But if an established enamelist sells the same ash tray to a collector of his works for $25.00, the student should not feel that he could ask more than $8.50, the recognized market price. The student should neither undercharge nor overcharge. Very likely, a student's work that is the result of serious study and instructor's supervision is far superior to that on the commercial market. By and large, a student should expect to receive about one-half of the price that his professor receives, as long as he maintains the status of student.

When it comes to prizes in any exhibition, certainly there should be no discrimination as to age, experience, or status. This is a matter which is and should be left indisputably up to the opinion of the judges. Retail stores demand commissions for selling ranging from 25 per cent up to 50 per cent. The latter percentage is more usual than the former. In order to make his time worth-while the student, if he wishes to sell to stores, must learn to simplify his work but not resort to gift-shop merchandise. He must minimize the number of firings and methodize each step of the work from preparing the copper to finishing the edges. Superficial patterns in gold luster or meaningless scattering of lumps do find their way to many store counters, but selling such work is hardly the way to establish one's reputation as a fine enamelist.

"BUTTERFLIES AND THISTLE" by the author; an example of 26-gauge round wire cloisonne; the enamel was not refired after final stoning and polishing, and the fine silver wires were subsequently gold plated. (Courtesy George Gund Collection)

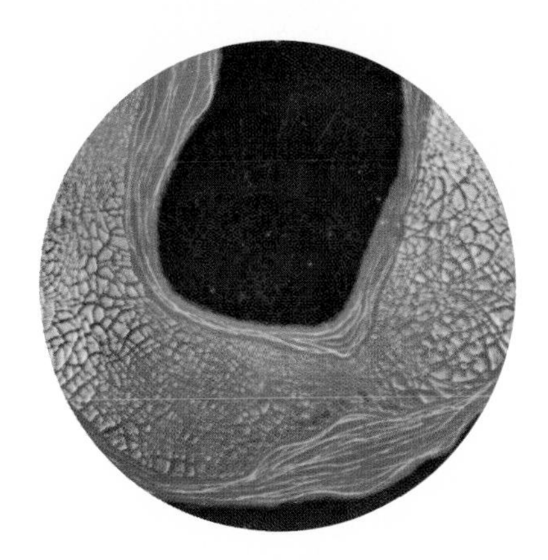

"SOMBER DEPTHS WITHIN" by the author; gold luster crackle on opaque red enamel. (Courtesy William Barrett)

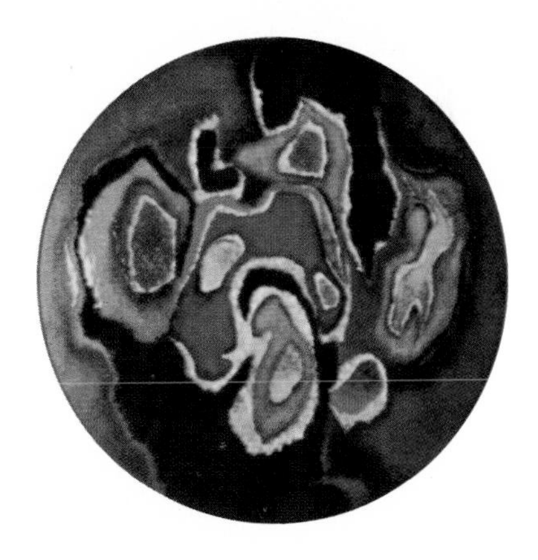

"OUTER SPACES" by the author; a 12-inch plate with wet-inlay technique. (Courtesy H. O. Mierke Collection)

CHAPTER 3

TESTS

The tests that follow require the consideration of every serious student of enameling. Making tests can be misconstrued as mere busy work, something to occupy the craftsmen's time on a rainy afternoon, and to be sure, one could spend a large proportion of his time testing and cataloguing enamels of every description and evading the real issue—that of creating a finished object from the tests. However, it has been my observation that most students do not take enough time for testing. Indeed, there are so many exciting effects that can be achieved through testing that it is to be hoped the following suggestions will serve only as a beginning to further research.

It is assumed that the entire working palette of fifteen to twenty transparent enamels and about the same number of opaque enamels were tested in the usual way at the beginning of the course.

For testing, see pages 39 & 40. You should now go further with the study of tests. As a musician continues to practice his scales, so must the enamelist continue to make tests at the outset of each new creation.

The tests must be carefully labeled, preferably by lettering with India Ink on masking tape and securely fastened to the reverse side. Years later, you will pat yourself on the back for having taken time out to catalogue your experiments, thereby saving hours of repetitious testing when the job at hand has a deadline.

1. *Tests for Copper*

Copper is an inexpensive metal and the one most commonly used for enameling, so we will start with tests on copper.

For the transparent colors the copper must be of the ultimate refinement. Copper with any silver or brass alloy is disastrous. Trouble begins on the fourth or fifth firing in the form of blisters (being the alloy which is released to the surface of the metal by heat). These blisters may be pricked open and filled with grains of enamel, but unfortunately, more blisters will occur with subsequent firings. A kind of highly-refined copper (known as electrolytic) that is over 99 per cent pure copper is now available through reliable concerns dealing in "enameling copper." Roofing copper or etching copper is inadvisable as a base for enameling. For the most brilliant effects the copper should be properly cleaned and prepared. There are various procedures and short cuts, but all amount to about the same thing in the end. First the copper should be annealed. Bring the piece to a pinkish glow (about two minutes at 1500° F.) This procedure kills two birds with the same stone—it burns away any grease or oil that might have been employed by the spinner or in the factory during the process of rolling and it also creates a slight fireglaze. This, when removed by acid, carries with it certain impurities found on the surface of the metal. Clean thoroughly, if possible in a fresh mixture of nitric acid. Put the amount of water needed to cover the object in a large Pyrex bowl or cake tray and slowly add the acid until the mixture turns the copper to a pale pinkish color. It is better to adjust the strength of the acid solution to suit yourself than to follow the time-worn recipe of one part acid to four parts water. Sticks or rags left in the acid make it muddy and discolored. It then becomes unfit for enameling and should be changed for fresher acid. Proper acid cleaning is one of the secrets for brilliant enamel work. Bring up a brighter surface now on the copper by rubbing vigorously with No. 00 steel wool. If the surface is slightly oily from the steel wool, saliva can be used to alkalize it, or a little detergent such as Joy applied with a rag and then rinsed off completes the preparation of the copper for enameling.

Commercial houses use a bright dip. This tends to enhance the clarity of transparent enamels. Some enamelers use a special cleaning process in addition to the regular one described above for transparent enamels on copper. This is a kind of "yellow pickle" composed of nitric acid and common salt, in the proportions of acid, one part, and salt 2 per cent, or roughly speaking about a teaspoonful of salt to a cup of nitric acid. The mixture should be stirred with a stick or wooden spoon until the salt is completely dissolved. The object is then pickled for not more than half a minute to one minute until it has obtained a yellow glossy surface. Afterwards it is well rinsed in water, brushed and dried properly by means of a piece of cloth. This yellow pickle increases the brilliancy of transparent enamels after firing.

It might be stated here that in the case of large murals, or perhaps for hastily made opaque tests, no cleaning of any kind need be done. The enamel will adhere to the copper, but of course, the true effect of deep rich transparent colors will be jeopardized.

a. SOFT AND HARD ENAMELS. It is necessary to become familiar with the hardness and softness of each color. If, for instance, a complicated design is to be rendered making use of a large palette, and hard firing enamels were placed at the center of a plate or bowl and excessively soft enamels near the perimeter of the piece, the outside enamels would be entirely burned out before the enamels in the center were even fused. This is a common mistake that is easily rectified with a few preliminary tests. Place the various enamels, both hard and soft, in the kiln at the same temperature. Make note of those that fire at a low temperature and those that seem to take a much higher heat and a longer time to mature.

Hard enamels such as hard white or hard ivory are useful as under-

FIGURE 38. Soft red when juxtaposed against hard flux creates interesting edges.

coats, but it is well to know how they react on subsequent firings. For large plaques, if a heavy cold weight is placed on the hard enamel surface, cracks will occur. These are very difficult to eradicate. With soft enamels as a background or undercoat, the cracks are more apt to flow together again. Soft opaque enamels have a tendency to burn in at the edges, especially such colors as soft red or vermilion. This can be very effective, and it is possible to place the piece in the kiln several times, turning it each time, to control such burned-out edges.

Soft white, especially, has a tendency to become greenish when fired directly over copper. Such effects can be made use of and controlled by skillful heat controls. It is possible to mix hard and soft enamels to produce a medium fusing color if such a situation arises. You cannot, however, create a new color by grinding one shade with another unless both have the same fusing point. Most opaque reds are likely to be soft colors, and when mixed with pink or light blue opaque the result will not be a lavender, but merely a light blue with black specks. This is obvious as the opaque red particles will burn to black long before the harder opaque turquoise or blue mature.

b. LOW AND HIGH FIRING. High firing usually implies a temperature above that normally used to melt or bring to maturity any enamel frit. We therefore do not say that this or that degree Fahrenheit is for high firing because, as stated above, the hardness or softness of the enamel itself is the determining factor for all firing. 1500° F. is the temperatures of the kiln used for *most* enamels, but so much more can be accomplished when the enameler is willing to overfire, or in some cases underfire. To obtain a burned edge or interesting halation with a soft enamel you might overfire (that is, fire at a temperature of 1700° F.), and by the same token, you might underfire (1450° F.) a particularly hard color. Make a test of opaque reds and vermilions fired high for interesting effects, and then try applying a very hard white or hard ivory in narrow lines or patterns, firing it only to the orange peel stage. This will produce an effect somewhat similar to Wedgwood pottery with raised or modeled lines or areas.

Dark opaque reds become darker with high firing. Most opaque blues and blue-greens show considerable variation according to the heat applied. Tests should be recorded of the enamels showing both normal, low, and high firing.

Transparent enamels seem to become brighter and clearer with numerous firings. Middle transparent blues and olive greens increase in intensity with subsequent firings. A slight application of enamel is added each time, and as many as seven or eight high firings will bring out effects with transparent blues that are very deep and rich in color.

C. TRANSPARENT ENAMELS ON COPPER. Transparent enamels on copper need not be dull and uninteresting. The beginning student is apt to make the mistake of loading the enamel much too thickly, with the result that, when fired at the average temperature of 1500° F., a dark semi-opaque effect is the disappointing reward for his labors. By applying a few basic rules the result should be quite different. For transparent enamels the basic rules consist of (1) proper cleaning and preparation of the copper, (2) washing and regrinding of enamels, (3) thin applications of the enamel, (4) high-enough firing, and (5) subsequent, or multiple, firings of the same enamel.

When to use a flux base for transparent colors and when not to is a personal choice. Many kinds of fluxes are obtainable from very soft fusing flux to hard flux, and special fluxes for red, gold, silver, etc. It is common knowledge that most transparent reds and pinks on copper

FIGURE 39. Raised effects like those in Wedgwood pottery are attained by underfiring areas of hard opaque white.

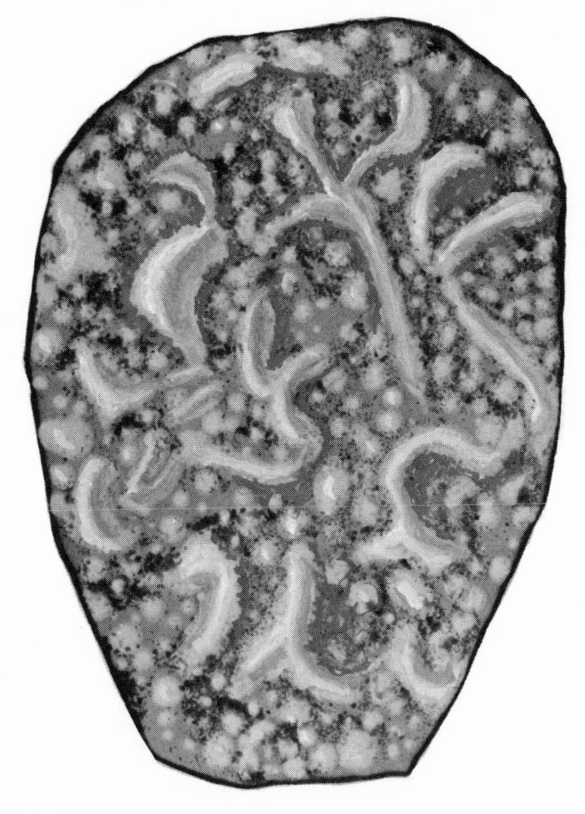

are only successful over a fired coat of flux, but brilliant effects can also be achieved with transparent yellows, golds, chartreuses, citrons, and tans when fired over one glowing coat of flux. The flux undercoater should be applied thinly and if possible slightly overfired. You soon learn by coloration when the flux is at its maximum brilliancy. In fact, if the kiln is without a pyrometer, or if you are firing by blow torch or with the oxy-acetylene tank, the coloration of the fused flux becomes a criteria for heat. A duller pinkish color of the flux denotes a lower heat, and a more orangey transparent color suggests that the temperature is at least 1500° F. to 1600° F.

First make tests of all opaque colors as they are obtainable from the factory. Next proceed to apply various layers of the same opaque color for better definition, and then try flux over the opaque, a combination that produces rather interesting textural effects in some cases.

Continue with all sorts of combinations of lighter opaques with thin coats of transparent enamels over them. Strange and interesting colors result from transparent lavenders over opaque greens, transparent ruby red over opaque red, transparent light tan over opaque yellow, transparent olive green over opaque blue, and so on.

FIGURE 40. Flux over opaque.

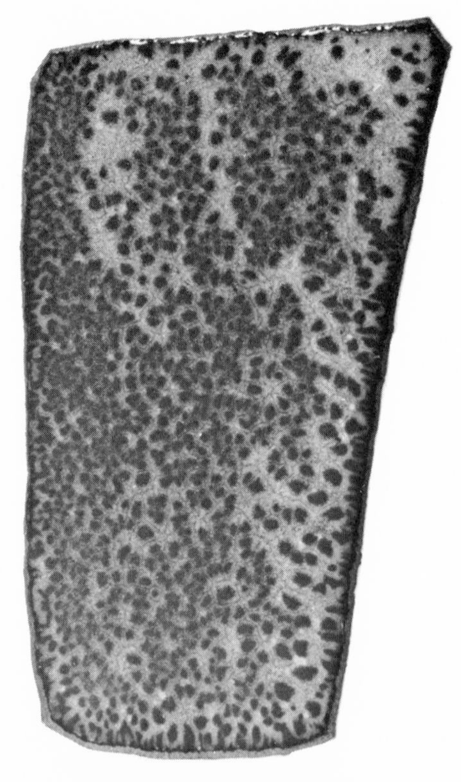

Now try the admixture of opaques with opaques. For this the colors must be ground together with a mortar and pestle. For instance, if you need a flesh tone, or perhaps a special color to dominate a mural so that the mural would fit the décor of a certain room, a series of proportionate tests would be made. Use a tiny measuring spoon or a small bottle cap, and start, in the case of a flesh tone, with measured proportions of perhaps two parts opaque white, two parts opaque pink, one part opaque tan, and one-half part opaque light yellow. The colors are placed in the mortar and ground to an extremely fine state. They should have the consistency of ground cinnamon or ground cloves. If the enamels were of the same hardness, the fired result would be quite similar to the color of the unfired grains when wet. However, seldom do all opaque colors have the same fusing point, so after keeping a careful record of the proportions used of each color in the mixture, simply proceed from there with further tests of varying proportions until the desired shade is achieved.

Tests for silver foil should include a variety of enamels upon which the foil is fired. Best effects are gained when transparent yellow of medium hardness is used as a base. Over soft flux the foil may show a more crinkled quality. A rather interesting gray color results from using a soft opaque red as a base. Sometimes the silver paillons that are placed upon a very soft red or soft white become quite dark and lose their brilliancy altogether. On hard opaque white, silver foil is definitely more successful.

If the undercoat of enamel is stoned after firing to release air pockets and then refired, the silver foil with its tiny pierced holes (about 100 per square inch) should lie flat and show no pits or "hillocks" from below the surface.

Make exploratory tests of all the above possibilities for your own record.

Also, with gold foil, which is sometimes sold in a thinner gauge, it is possible to overlap the paillons or larger pieces. Hold the delicate gold foil between a double sheet of tracing paper with the fold at the bottom, designate the shapes needed, punch holes through the foil while placed on soft cardboard, and gently separate the paper from the gold without tearing it.

The gold may appear to turn black from slight overfiring, but transparency is still maintained after firing the enamels over it. When working on fine gold (24K) apply the enamel by the wet inlay method with utmost care and respect. Try to lay on the colors one grain high, fire, and repeat the process with another one-grain-high application. Subsequent applications may be different hues or hues blended together.

2. *Tests for Silver*

Tests for silver are not unlike those for enameling on silver foil, at least in respect to color effects. However, there is a more limpid quality, a more brilliant unbroken sheen because the surface of the silver remains smooth rather than crinkled as in the foil enameling.

a. STERLING SILVER. Sterling silver can be enameled with some success, but the colors are less intense than when applied over fine silver. Remember that sterling silver has a melting point considerably lower than copper, and also lower than fine silver. Because copper melts at 1981° F., fine silver at 1761° F., and sterling at 1640° F., some caution must be taken in regard to overfiring with sterling.

b. STERLING WITH FINE SILVER SURFACE. For small pieces of enameled jewelry it is usually better to work with fine silver. In some cases when hardness is essential, sterling silver (actually an alloy of 925 parts fine silver and 75 parts copper) being harder than fine silver, can be so treated as to bring the fine silver to the surface for more brilliant enameling, but still retaining the hardness of sterling silver. The process for doing this is as follows:

After giving the silver a good buffing with the felt wheel and then the wire brush, using a compound of oil and pumice, wash it with soap and a fine hand brush. At this point the silver can be polished with jeweler's rouge, using a cotton or muslin buff. Wash it again in warm soapy water. Now you are ready for the actual process of bringing the fine silver to the surface of the sterling. Step one: Heat to annealing temperature (set the kiln at 900° F. and let the silver remain in the kiln until you begin to detect a pinkish color on both the silver and the trivet). Step two: Remove from the kiln and air cool until it can be picked up with the bare hands. Step three: Pickle in a hot (not boiling) solution of sulphuric acid with the proportions of one acid to ten parts water. Keep it in the pickle for at least five minutes and then rinse thoroughly. Step four: The silver is now burnished. This can be done with a fine brass wire brush and plenty of soapy water, but use a slow speed on the motor. If the wire brush and slow motor are unobtainable, a very fine steel wool can be used with the soap and water as a burnishing process. Caution must be employed in rubbing with steel wool so that the fine silver is not worn away but only burnished. The entire process (steps 1, 2, 3, and 4) is repeated three times until dark areas have disappeared from the surface, and a sufficient deposit of pure silver is built up on the sterling silver base metal.

c. FINE SILVER WITH STEEL WOOLED SURFACE. There are several methods of preparing fine silver for enameling. The first is a scratched or satin finished surface. To do this first burn off any greasy or oily substance (which might have been on the silver as it came from the mill) by applying a rather large blowtorch flame to the silver. By working in a dimly lighted part of the studio, or by turning off the light, you can detect the pinkness of the silver at the edge of the metal. Do not heat above this stage of pinkness. Immerse when warm into the weak sulphuric acid pickle mentioned above. Leave it in the pickle for a few minutes and then rub vigorously in a rotating motion with fine (00) steel wool. Sometimes saliva is wiped on the surface to remove any grease that may have been deposited from the steel wool. Keep fingers from touching the surface and start the enameling. If tracing is necessary, first use a red carbon paper and fine pointer or sharp pointed 6H pencil, then rescribe the carbon line directly on the silver, after which the cleaning process must be repeated. All silver enameling—for that matter, *all* enameling—over silver paillons should be carried out in the most meticulous and exacting manner. Only the brilliantly transparent colors are acceptable and they should be applied evenly in a thin layer *one grain high.* If, after the first firing there are open areas, these disappear with the second application.

d. FINE SILVER WITH POLISHED SURFACE. Another surface for fine silver is one that produces greater brilliancy and a kind of shiny or glassy quality to the enamel. Perhaps, in some cases, this effect is too intense and if used to excess might be of questionable taste. Enamels sometimes tend to look too much like Christmas tree ornaments, or anodized aluminum, instead of having the deep, limpid quality so unique to this particular medium. However, the process of achieving a polished surface is quite simple. It consists of annealing and cleaning in the sulphuric acid pickle as mentioned above, but followed first by buffing and then by polishing. Note that if counter-enameling is done the annealing process would naturally be eliminated. To buff the fine silver for enameling first use the felt buffer on a medium speed motor. Bobbing compound, followed by tripoli and sometimes white diamond rouge complete the buffing process. Muslin buffing wheels with the tripoli may be employed after the felt wheel but will perform less efficiently for removing tiny scratches. Remove any compound left on the surface of the silver by wiping with a dry rag; if the compound is stubborn, dip a soft nylon-bristled toothbrush in household ammonia and rub it on a cake of Ivory soap for thorough cleansing of the silver.

Bring the fine silver to a still brighter surface by polishing. For this, a soft cotton buff with jeweler's red rouge is gently applied with little pressure to bring out a mirrorlike glow. Enamels over such a buffed and polished surface have greater intensity than over a satin-scratched or steel-wooled surface.

e. FINE SILVER WITH VIBRATED SURFACE. Your tests for fine silver should include one more experiment—that of the vibrated or textured surface. The discussion of the work of Carl Fabergé in Chapter 1 mentioned that he was noted for his experimentation and development of the machine-turned surface called guilloche patterns (spiraling and rotating of the engraved metal showing through the highly translucent enamels to produce a special kind of brilliancy). Perhaps, for some, this effect is too mechanical or commercial-looking, but Fabergé's enamels have seldom been equaled and have become collector's items of considerable value.

Most craft-tool suppliers carry what is called a vibra-tool, and with this tool it is possible to approximate the machine-turned surface popular in England at the turn of the century and still widely used for commercial name plates, insignias, lodge pins, etc. Hold the vibra-tool at right angles to the surface of the metal with a medium fine point. With a little practice and control you may choose among the following patterns: series of close vertical lines; cross-hatching of vertical lines; stipling or dots at close intervals; rotating in circular motion; an all-over effect using the tool in all directions to texturize the surface completely; and a kind of pattern made by vibrating the surface with a combination of dots, lines, and circles in regular repeat.

Cleaning of the silver is unnecessary after the vibrating as the metal is quite "raw" and receives the enameling very readily. Try a test where the vibrated surface is combined with a plain surface that has previously been polished.

3. *Tests for Sgraffito*

The term "sgraffito" is a craft word in common use although many dictionaries do not explain it. It means the scratching through of an outer coat of paint, glaze, or enamel to expose or delineate a design on an undercoat. So many effects are possible with this technique and it provides such an excellent opportunity for the craftsman to utilize his skills as a draftsman that time should be taken to test many examples for future reference. Of course, you may arrive at your own manner and style of sgraffito, but there are a few essentials to be noted if success is to be assured. First, a certain stage of moistness must be explored. If the enamel to be sgraffitoed is too moist, the line merely floods back into its own space leaving no designation at all. If the material is too dry, the grains fall away, often creating too broad a track made by the tool. Secondly, it is best to sift on the enamel through a rather fine mesh. A mesh of 100 wires to the square inch is better than the more often-used 80 mesh screen. Note: You can make your own fine meshed screen by stretching a piece of nylon stocking over an improvised frame. Wet the surface of the metal first, using a weak solution of tragacanth in the atomizer, then dust the dry enamel onto the moistened surface, and continue to maintain a state of dampness if large areas dry out.

For the pointer, here again it is a matter of choice, but you will find that if your degree of moisture is just right, and if the enamel is finely dusted, a very fine jeweler's scribe or steel pointer will work to

FIGURE 41. Sgraffito one coat of flux on copper, clean surface, and apply a second coat of flux.

perfection. I find that the favorite pointer is a sharpened end of a small water-color brush handle. This is a resilient point and can be adjusted as the occasion arises. Scratch the line and gently blow the excess away. (The enamel should stay in place because of the tragacanth spray mentioned above.) Clogging at the edge of the line can be avoided by using a thinner coat of enamel, unless such an effect is desired. It is not the intention of the sgraffito technique to look hard and stiff-edged, but to have a character of its own. Making use of broken, shaky, or textured lines can be much more effective than hard lines. After such a line is fired, a certain amount of freedom and looseness will be enhanced.

a. TRANSPARENT TO OPAQUE. First the tests should include a very simple type of sgraffito—the opaque base with transparent enamel superimposed and scratched to show the under color. To do this start with fired coats of opaque of light values either in soft or hard enamels. It is obvious that to reverse this (light transparents on dark opaques) very little would be gained. Some of the more effective experiments that you might try are: a transparent medium tan or dark mahogany brown on light hard opaque ivory; transparent dark lapis or cobalt

FIGURE 42. Sgraffito with light transparent to copper in broken areas, by Virginia Gaz.

FIGURE 43. Sgraffito dark transparents on light opaques.

FIGURE 44. Sgraffito of flux to copper with flux applied to whole tray, by Nancy Brown.

over medium opaque turquoise or opaque robin's egg blue; transparent dark claret or wine red on opaque vermilion is apparently successful as is transparent ruby red used over the opaque pinks. By using an extra hard opaque white as a base and sgraffitoing through transparent ruby red or maroon with a very fine line, an etching or wood-block quality is possible.

b. OPAQUE TO TRANSPARENT. Now the process may be reversed. First fire a clear coat of medium or soft fusing flux on the copper.
Proceed by dusting opaque black over the entire surface and sgraffito through to the flux. The line when fired high appears to be almost golden and is worth the experiment. Try the same process but supplant the opaque black with opaque white, red, or tobacco brown for unusual qualities. One more—the use of opaque medium blue (delft blue) on the transparent gray with the sgraffitoed line is most appealing.

FIGURE 45. Sgraffito light opaque over dark transparent.

C. BLACK CRACKLE SGRAFFITO ON FLUX. A product of more recent development is that known as crackle enamel. This is a soft slush type of inexpensive enamel that comes in many colors. It has little of the true qualities and characteristics of fine translucent enamels and, if overly used, is of questionable esthetic significance. However, there are times when crackle can be effective, especially when used as an undercoat or perhaps as a sgraffito to be later covered with transparent enamels. First, fire an even coat of medium fusing flux and then apply the black crackle. This is done with a water-color brush as the crackle is the consistency of tempera paint. Let the coat of black crackle dry, having applied it as evenly as possible, and not too thickly. Sgraffito your pattern or line arrangement with a very sharp steel pointer and fire at about 1550° F. or slightly over. Your sgraffitoed pattern will be distinct and show in contrast to the supplementary crackles that are inevitable. High firing will sometimes sublimate an otherwise commonplace motif to a rather amazing organic pattern. White crackle over a soft fusing flux can also be sgraffitoed to produce a pleasing quality, but much is left up to chance in this kind of work. However, it is worth a try especially if the "crackle-sgraffito" is used only as a kind of underpainting for textural effects, much as in oil painting.

FIGURE 46. Sgraffito of flux to copper showing burned out line, by Josephine McCorckle.

d. BLACK OVERGLAZE SGRAFFITO ON ENAMEL. The next two experiments for tests involve the use of black overglaze. Black overglaze is, in effect, nothing more than an extremely fine-ground enamel, identical with certain forms of old-fashioned china paint. (Some of the small vials of black powdered china paint resurrected from old attics are rather precious to the enameler who wishes to experiment. In fact, they constitute a kind of quality that is not obtainable today.) To use the black overglaze or some of the black "outliner" china paint color, grind the pigment on a piece of glass or perhaps a ceramic tile. Do this with an oil-painting palette knife, using a drop or two of oil of sassafrass or oil of lavender to extend the solution, making it into a consistency that will flow readily from a brush or pen. A drop of squeegee oil is also added to act as a binder.

The first experiment consists of painting a very even coat of this black overglaze, in the consistency mentioned above, directly on a clean plate of copper. After drying, it will be possible to sgraffito lines, and areas, or to remove the entire background surrounding certain motifs. Now, before firing the experiment, dust a thin but even coat of flux over the surface. Notice that the black overglaze lines and patterns

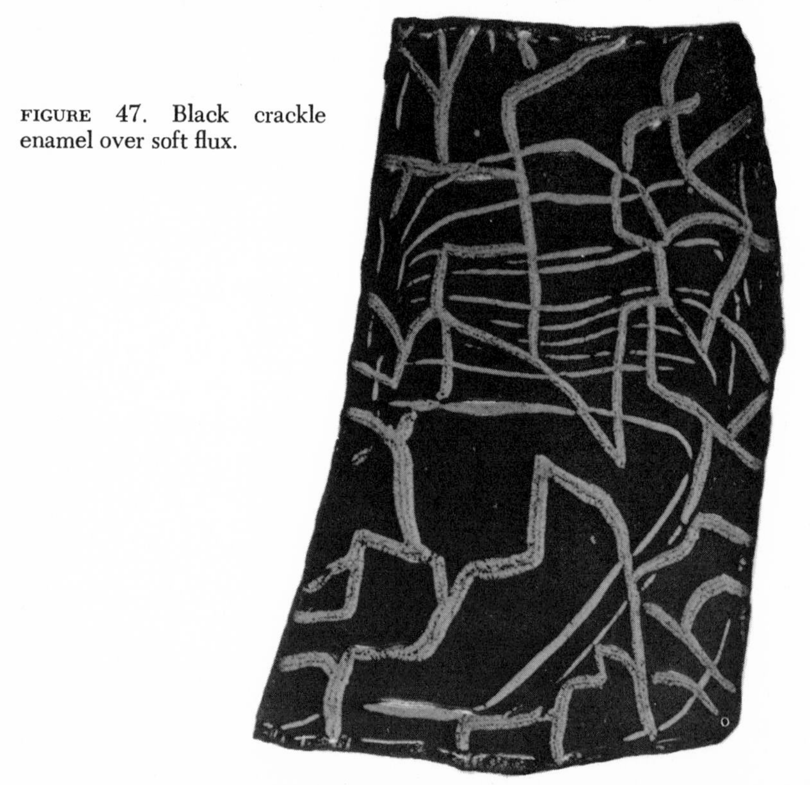

FIGURE 47. Black crackle enamel over soft flux.

FIGURE 49. "Rooster" by the author; black overglaze line painted directly on copper. (*The Cleveland Museum of Art, lent by the artist for the forty-sixth May Show*)

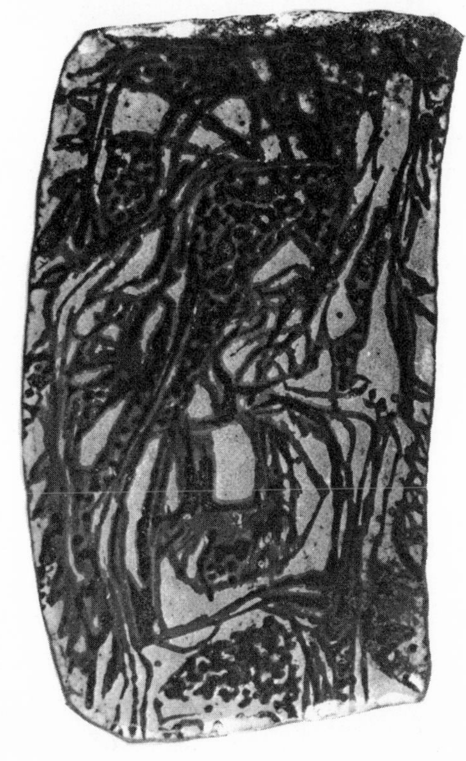

FIGURE 48. Black overglaze directly on copper, sgraffito light areas.

remain exactly where placed and, what is more to advantage, show no burning-out or "sinking" quality so common to overglaze when fired high. The finest of pen lines, and the most meticulous sgraffito work with hair-thin lines can be carried out in this manner.

e. BLACK OVERGLAZE SGRAFFITO ON COLORED ENAMEL. In this test fire a coat of enamel on the copper *first,* after which the black overglaze should be applied in an even coat all over and then sgraffiitoed for the desired pattern or line design. This experiment, you will notice, is the reverse of the preceding experiment. There is the advantage of obtaining a different background color (sometimes a white background is desirable) but also a slight disadvantage—when the overglaze is fired at a temperature high enough to make it sink or "eat in" to the undercoat. At all times when firing overglazes the process must be watched with utmost care. It is well to place the piece near the door of

FIGURE 50. Black overglaze on enameled surface, plus sgraffito line.

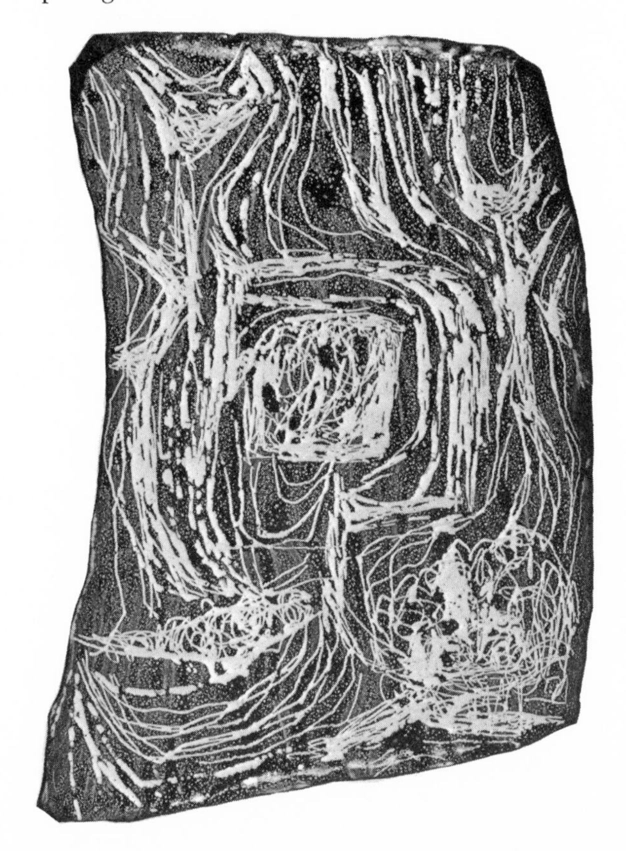

the kiln and keep the door slightly ajar to detect the very moment of maturation of the overglaze. A delay of only a few seconds will cause the overglaze to sink into the surface, leaving a depressed line.

f. FLUX SGRAFFITO ON COPPER. You might try one more kind of sgraffito work. This is a most useful technique for producing an organic or natural line. Dust a fine coat of flux evenly over the surface of cleaned copper. With the flux slightly damp, make the sgraffito line directly through to the copper. Fire at the usual temperature, (1500° F.). When this is accomplished, a very interesting type of line is produced by the exposed oxidized copper. Sometimes with a broad line there occurs an oxidized edge on both sides of the line. With fine steel wool, or a mild solution of nitric acid (five parts water to one part acid), clean the oxidization away and apply another coat of flux, bringing this to a higher firing (1600° F.). This process is similar to one which uses a commercial product known as Klyr Cote and which will be discussed later in Chapter 4 under techniques.

4. *Tests for Plating*

a. GOLD AND SILVER PLATING. Silver and gold plating are not as expensive as normally thought. So often the change of color to the exposed metal parts of an enameled piece not only enhance its intrinsic value but also its esthetic qualities. Copper, which can be plated with gold, silver, or chromium, lends itself to many more enamel colors and relationships. There are innumerable colors of gold, including white golds, yellow golds, red golds, and green golds. Some red golds are very close to the color of polished copper, and some yellow golds appear very much like brass. Consequently it is necessary to specify what color is desired when you present your piece to the plater. Edges of bowls can be successfully plated without danger of cracking the enamel if there is no particular tension or stress involved by the use of too wide a variety of soft and hard enamels. Remember that the plating process does imply some heat, but, what is even more serious, the piece must be immersed in a series of acid baths. These acids are not detrimental to most enamels, but if bezels, pits, or cracks are present where the acid can penetrate or remain, there is sure to be trouble. The plater usually polishes the piece after plating, and therefore it is always advisable to ask for double plating. The original polishing, of course, must be impeccable. The tiniest scratch will become quite evident after plating. Edges to enameled bowls made by placing a U-shaped fine silver capping can be gold plated for a rather elegant effect.

b. PLATING FOR CLOISONNÉ. Plating, either gold or silver, for cloisonné requires a special consideration: whereas the procedure in plating consists of attaching a piece of pure gold or silver in the form of a wire or small tab to the metal to be affected, and whereas these metals must come in direct contact, it follows that all of the cloisonné wires must touch each other or be soldered to the base. With ribbon wire this is somewhat less difficult than with round or square cloisonné wire. However, sometimes the fact that here and there the wires do not take the plating may create an unusual play of color. Heavy plating of gold negates, to a great extent, the possibility of further tarnishing or oxidization.

c. PLATING FOR CHAMPLEVÉ. Gold or silver plating is most practical and seems to be more expedient in the case of champlevé. In this

FIGURE 51. "Lichens" by the author; champlevé enameling with gold plating, the design of which was derived from the pattern left by fire-scale patches. (*Courtesy Barry Bradley*)

technique the enamelist is primarily concerned with the play of the metal color in juxtaposition with the chosen enamel colors.

If he is considering cool blues and greens against a white metal, he is still able to use copper as his base metal and plate it with silver. This may not need to be done too often as the cost of silver is not exorbitant, and most craftsmen would prefer to work directly on silver for champlevé designs. But there might be cases where larger areas were needed, making copper a more practical metal to use, and also there are certain tones of transparent enamels that are obtainable only over copper. Gold plating can be done on either silver or copper and is effective when combined with warm transparent enamels. The champlevé technique will be described in detail later (Chapter 6) but at present we are concerned mostly in how to prepare the finished champlevé for plating. There is the problem of first removing all of the fire glaze from the exposed metal after which the most minute

FIGURE 52. Break-through with one color.

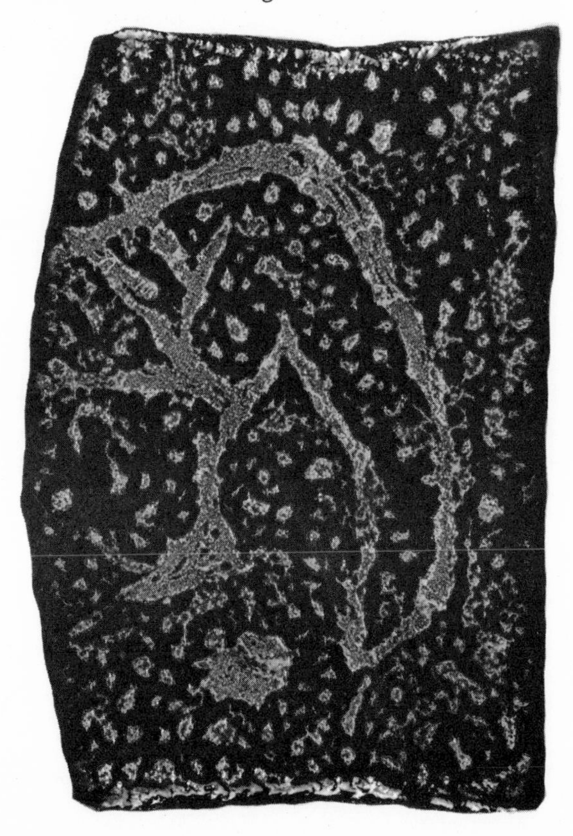

scratches must be eradicated before plating is considered. To do this, arduous stoning *under water* is first performed to bring the enameled areas to the level of the metal. First use a coarse carborundum stone working in a circular motion, and follow this with a medium stone, next a fine stone, and lastly a Scotch hone. Start the buffing with a felt buff and tripoli. Keep the piece moving in a circular motion; otherwise you will discover that you have worn away the metal by buffing leaving the enameled areas raised. (Note: Enamel is harder than most metals.) This sunken quality to the champlevé is distressing and unprofessional. Also, when buffing in one direction you may find by close scrutiny that there are tiny "fish tails" at the edge of each enameled area. This can only be avoided by restoning and buffing again in a rotary motion. It is rare after final polishing not to detect a few tiny pits in the metal. These are caused by minute air bubbles in the asphaltum resist, which allowed the acid to penetrate. To eradicate the pits two suggestions might be given: One, apply a second thin coat of asphaltum; two, use a weaker acid and a longer period of immersion. When the champlevé piece is completed with no scratches or mars of any kind on the exposed metal areas, it is ready for plating.

If you are combining cloisonné designs within the recessed areas, the polishing process would be equally meticulous, and here again, for plating the wires must touch the metal itself and either be contiguous or otherwise soldered to the base. It would be wise to make several of the tests for plating mentioned above before proceeding to a more involved enameled piece.

5. *Tests for Break-Through*

a. ONE TRANSPARENT ON SOFT OPAQUE. Included in your tests should be a correctly labeled list of tabs showing the effect of soft enamels that tend to "break through" the surface enamels. A break-through will happen many times accidentally, but the enamelist should take time to tabulate a few of these for future use. Start with counter-enamel for these tests because the upper surface may require several coats. For the first coat, fire a fairly heavy layer of the softest opaque chartreuse. The next firing could be transparent medium turquoise. Raise the temperature to about 1600° F., and the chartreuse will break through the surface of the turquoise and at the same time become dark green or slightly transparent at the edge. There are certain opaque soft turquoise and gray enamels that become decidedly transparent with high firing and, strangely enough, will resume their opacity if refired at a lower temperature. Try the transparents also on soft fusing white in the same way. Such an effect is useful at the edge of a bowl or plate. Remember the procedure is always a light soft opaque *under* a dark or middle transparent. This "spotty" or "feathered" look is produced only when the two enamels are fired at the same time.

When the undercoat is fired separately and the darker transparent is dusted in preparation for the second firing, moisten the plate, dust, and then, by holding the atomizer very close to the enamel, purposely give it a few sudden squirts, which will produce a fascinating series of tiny circular spots. This texture will add to the break-through texture when fired.

An interesting break-through test, which might otherwise be listed under a kind of crackle, is achieved when an entire coat of soft opaque vermilion is fired at 1450° F., and then a very even coat of gold luster or other metallic luster is painted over the entire surface with the small brush in tiny strokes. Fire again and a most fascinating pattern of vermilion will break through the gold.

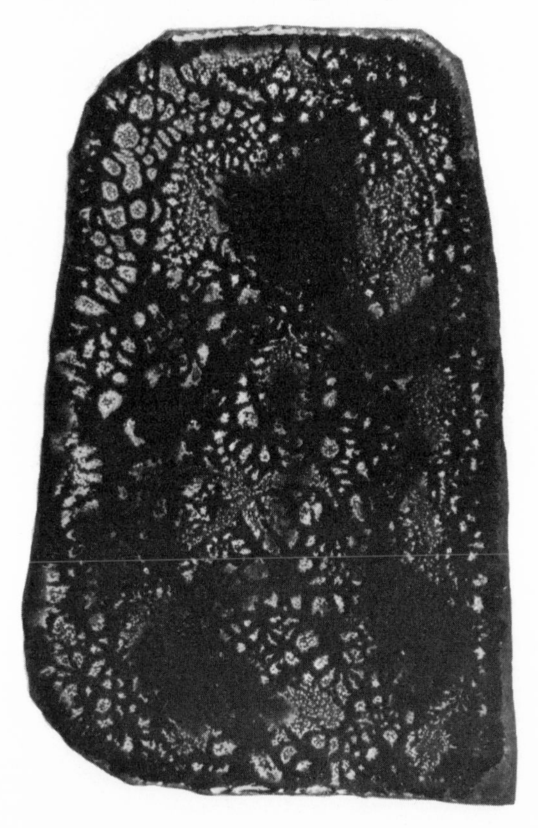

FIGURE 53. Break-through with more than one color.

b. TWO OR MORE TRANSPARENTS ON SOFT OPAQUE. Break-through effects are richer and more colorful with two or more transparents. To achieve these will require five separate firings.

Firing number one will be the counter-enamel to safeguard against cracking off of the four coats on the reverse side. Firing number two is a clear flux fired at average temperature. Proceed now (firing number three) with a light soft opaque, being sure, by previous testing that it is one of your softest opaques. Firing number four should be any light or medium transparent. The fifth and final firing can be a darker transparent color, or even a darker opaque color, but this time raise the temperature to 1600° F., or more, to cause the break-through of the several colors. Try many combinations for your tests, but be sure to make correct notations as the effect of overfiring may cause strange and intriguing color changes.

CHAPTER 4

TECHNIQUES

Before creating a design for enameling, it is important to choose the technique that will best express your idea. Certainly a motif or composition to be rendered in cloisonné would be designed in a different manner than one for champlevé or direct wet inlay. One type of design implies outlines as in cloisonné; for champlevé it is the play of metal areas against enameled areas; another technique presupposes solid color, with no concern for the exposed metal.

I should like to list and describe some of the better-known techniques for enameling, excluding cloisonné, champlevé and plique-à-jour, which will be treated in detail under separate chapters.

1. *Spatula (wet inlay)*

Spatula work, sometimes referred to as "wet inlay" is the most basic and best known of techniques. It is the method used to introduce the student to the craft, and yet, at times, it is so poorly carried out, or done in such a desultory manner that perhaps a few suggestions here might help to raise the standards. In some European countries wet inlay work is done with a small brush; however, with enamel that is generally sold ground as coarse as 80 mesh, this becomes a very slow and tedious method. The most expedient manner for working with wet inlay is to make use of two tools—a flat or spoonlike tool held in the left hand and a pointed, bent wirelike tool held in the right hand. Use the left hand to pick up the enamel while moist from a small jar or dish and deposit it on the metal surface; push or level the enamel into place with the right-hand tool. A common mistake—and one that must be avoided—is inlaying while the enamel is too dry. Spray the design continually using a mixture of tragacanth and water in the atomizer. The correct amount of moisture (near the saturation point) makes the tiny grains flood together and also enables the craftsman to juxtapose as many colors as he wishes. Soft edges or hard edges are controlled by the spreader. All areas are kept at the same level—this is important.

After firing is completed, there should be no bubbles or pits. When bubbles or blisters do occur, one of several reasons may have been the cause: (1) inferior copper, or copper in which there is too high a percentage of brass (never enamel on anything but the purest copper); (2) improper cleaning of the copper caused by dirty, overused acid, or greasy surfaces that were not eradicated by preannealing, or careless finger prints; (3) trapping of air pockets or bubbles beneath the surface before firing. When the enamel is being rewetted constantly, as in the case of an elaborate design, it is never advisable to rewet one section only or to use a too-concentrated mixture of tragacanth in the atomizer. (This will tend not only to trap bubbles beneath the encrusted surface but inevitably will cause discolored areas, especially when ordinary

FIGURE 54. For spatula work the two tools are held close to the inlaid design.

tap water is used instead of distilled water); (4) improper washing of the enamel. Enamels should be washed and reground in the mortar and pestle with each new design, even though there has been a thorough washing previously. Particles of impurity and dust from the air should be removed by grinding and washing. Pour away all cloudiness, or precipitate when washing until the water is absolutely clear. Acidulating by adding four or five drops of nitric acid, and then rinsing thoroughly with water several times guarantees further clarity and dissolves metal particles and foreign matter.

With the spatula method, shaded and textured areas may be built up with subsequent layers. When these are applied, always build up the entire piece including the background to the same level. If the original color or shade is desired, flux is used in lieu of any additional color for final surfacing.

A good recipe for gum tragacanth is to place one teaspoonful of the powdered gum in a quart mason jar, add one teaspoonful of pure wood alcohol, shake vigorously until the gum is dissolved, and then, while still shaking, slowly fill the jar with water. Allow this mixture to

FIGURE 55. Hand-hammered tray showing spatula technique, by Nancy Brown.

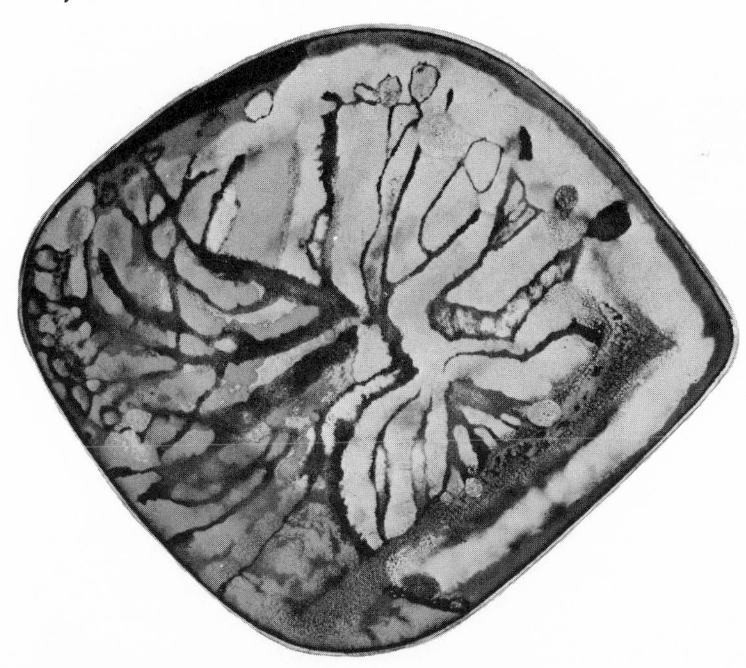

stand over night. Pour off the upper two-thirds of the quart for the atomizer solution (again diluting it two parts water to one part gum) and retain the thick creamy bottom third for special work such as cloisonné or champlevé on vertical sides of bowls. The gum solution is often used in a more concentrated state. (There is on the market a product known as Klyr Fyr, a convenient, ready-mixed liquid.)

When enameling a piece with twenty or more colors, place each enamel in a separate small porcelain dish or compartmented palette (plastic egg containers or plastic covers are excellent); then add some of the concentrated gum tragacanth solution to each one. In this method the enamels are used in the consistency of thick tempera, each one being easily accessible to the craftsman. This is a most expeditious manner of working.

Much European work shows textured and sometimes multicolored inlaid surfaces. These are achieved by applying small drops of slightly contrasting colors on the surface at close intervals while the enamel is still wet. After firing, the surface is carefully stoned, thereby giving some variation to the otherwise flat and uninterestingly plain areas. This is especially true if transparent colors are used, which tend to blend and fuse into the original coat of enamel.

FIGURE 56. Abstract plate by the author; colors dusted on rather than applied by spatula; break-through technique gives the final effect.

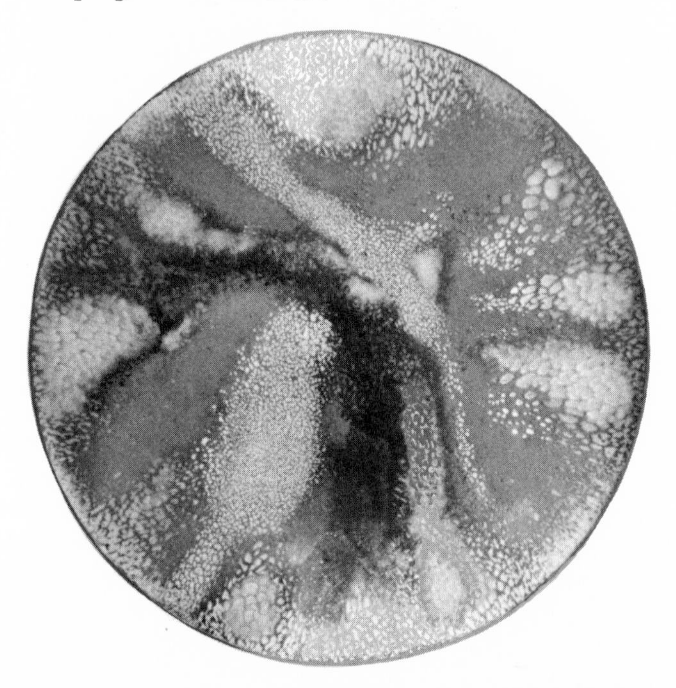

2. *Dusting*

Dusting and spraying of enamels has its place for larger pieces, backgrounds, and production work, but the spatula (or wet inlay) method still remains the basic technique for more intricate and precious pieces of enamel.

For counter-enamel and for plain backgrounds, or undercoats such a flux or hard white, the dusting method is the most logical to use. Although you can buy dusting implements, including tea strainers and gadgets of every description, most craftsmen, prefer to improvise their own dusting screens. You might start by making a small basket with ¾″ high edges folded up from a 4″ square of copper or brass wire mesh. For general dusting an 80 mesh wire is best but for more fine dusting use 100 mesh wire. To your collection of dusters, add frames with various grades of nylon mesh with cheesecloth stretched over them for extra fine dusting. Discarded tempera bottles and tiny pill bottles can be used as detail dusters. Cut away the metal cap *close* to the rim of the bottle and insert a disk of wire mesh. With practice, it is

FIGURE 57. Tray with colors dusted over fireglaze patches, by Nancy Brown.

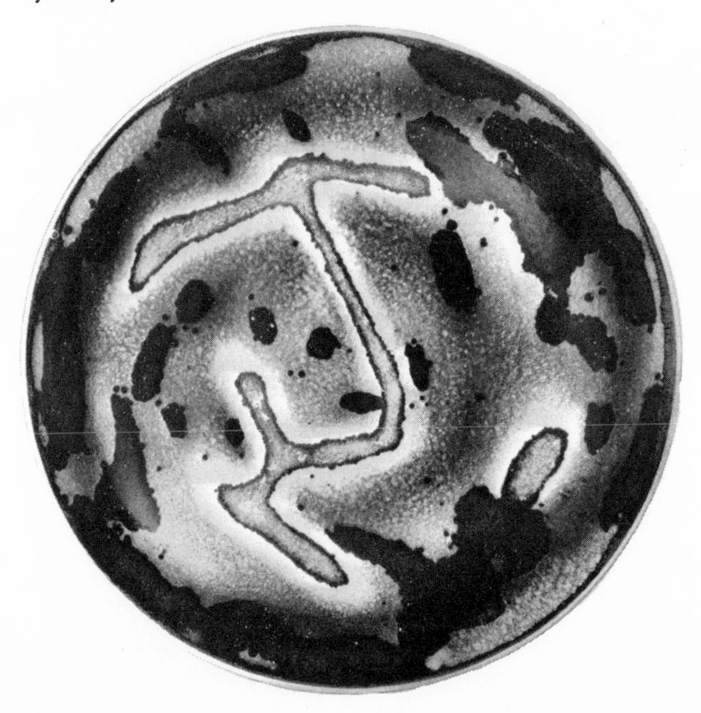

possible to dust a thin line of enamel on a moist surface and also to control the dusting for smooth shading.

Only by dusting the washed and dried enamels can you expect the maximum of transparency to the color. It is wise not to dust too heavily except with opaque enamels. Completely wetting and rewetting the surface of the enamel to the saturation point when dusting is a method used especially if the piece to be enameled has vertical sides.

To be sure, dry dusting is perfectly possible on a flat surface or one that is slightly bent—and actually there is less danger of bubbles forming when the dry method is used. By the addition of several colors fired at the same time (page 87), the effect of dusting becomes much less mechanical.

FIGURE 58. "Image and Constellation" by Mary Ellen McDermott; dusting combined with sgraffito. (*American Craftsmen's Council, New York*)

3. *Dropping*

Because enameling techniques are fascinating, too often they become the sole preoccupation of the craftsman. But, on the other hand, if an artist has a large repertoire of techniques upon which he may draw, he is that much more capable of expressing a variety of ideas.

One effective way of working that produces a quality unlike any other (colors blended abstractly, with very soft edges) is to discard the spatula and dusting basket altogether. Place several clean white papers on the bench and pour heaps of selected dry enamel on them. On a fired base of flux or almost any background color, spray or brush the surface with gum tragacanth. Now, take pinches of the dry enamels in your forefinger and thumb and try dropping them or spreading them on the surface. This becomes much like painting with colored sand and can produce most expressive results.

FIGURE 59. "Space Objects" by the author; an example of dropping the enamel when in a dry powdered form. (*The H. O. Mierke Collection*)

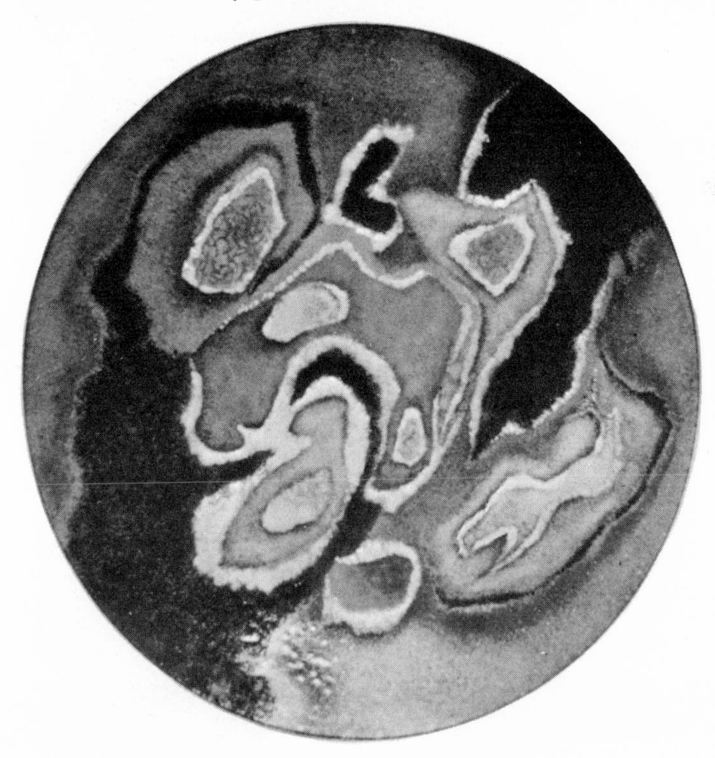

Another experiment is to wet the dry enamel after dropping it from the fingers, and by merely touching the surface with the tip of the finger, hazy and diffused spots are created, resembling dandelion fuzz or stars on a cloudy night.

Dropping blobs of moist enamel, splattering, and splashing are other possibilities.

4. *Brushing and Dump-off*

The term "dump-off" is used to describe a kind of technique that assures a completely free and organic character to enameling. For the

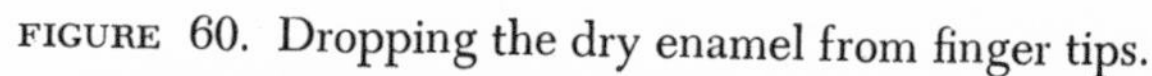

FIGURE 60. Dropping the dry enamel from finger tips.

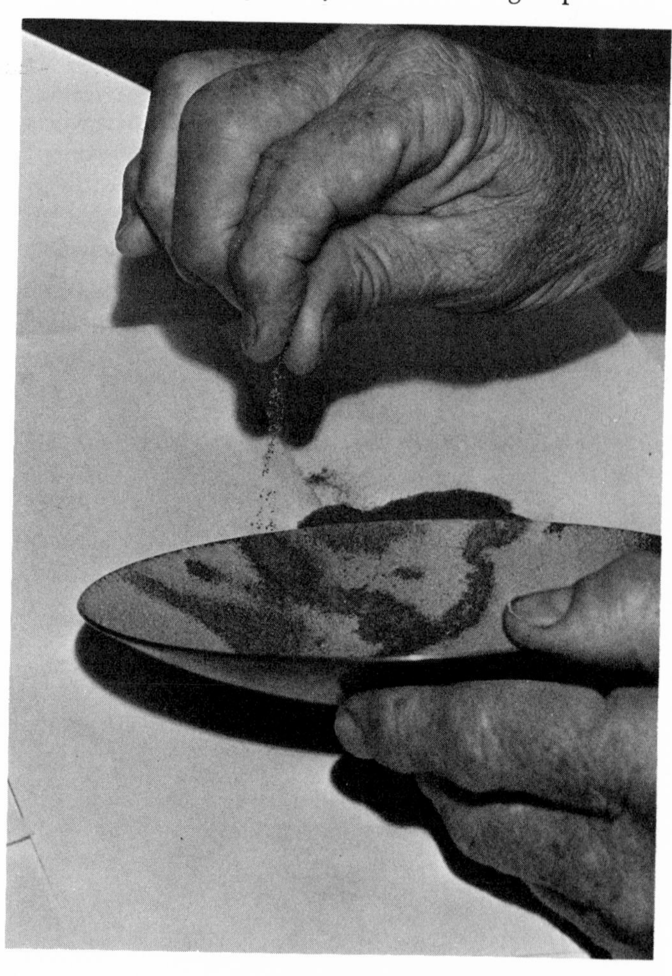

first experiment prepare a clean dry surface to the copper, and then with a fairly wide, flat water-color brush paint gum tragacanth solution in a broad abstract design. Notice that the edges crawl (leaving the surface of the copper slightly greasy creates a still more "crawly" effect). Now, with the coarse dusting basket quickly dust a coat of dry enamel over the entire surface. Hit the edge of the bowl or plaque on the bench over white paper dumping off all of the excess enamel, allowing it to remain only where it adheres to the tragacanth.

This same technique may work even better with squeegee oil, or Klyr Fyr. If opaque enamel is used, any unwanted smudges should be removed. The next step is to fire at about 1550° F. The edges of the enameled areas are most interesting and free, especially when fired high. Proceed to clean the exposed copper areas until they are either entirely or partially free of fire glaze. Repeat the steps above and build the design with continued firings of transparent enamels. Finally, dust

FIGURE 61. Dump-off.

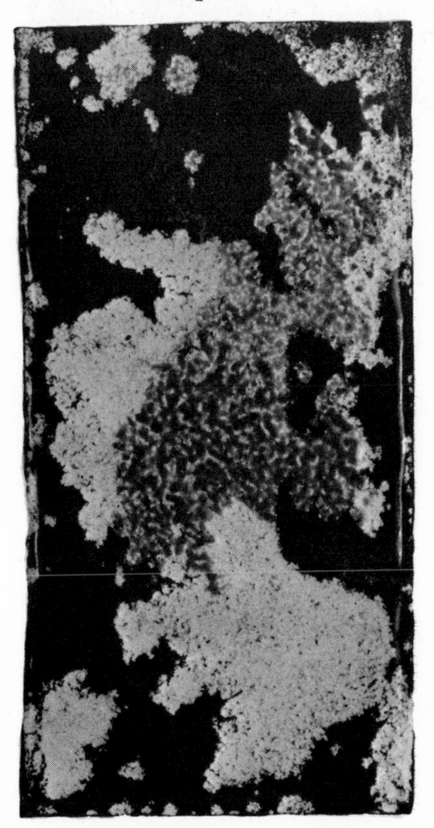

a coat of clear soft fusing flux evenly over the whole surface and fire at about 1600° F. This technique may be combined with a sgraffito line, allowing this line to have a similar organic and free character.

5. *Burn-out*

So often one discovers exciting, new, and useable techniques by accident. Sometimes "burned-out" pieces or "burned-up" pieces have a way of being more attractive than those to which such an accident had

FIGURE 62. "Christ with Angels" by the author; burn-out with silver cloisonné. (*The Cleveland Museum of Art, lent by the artist for the forty-fifth May Show*)

not happened. In the chapter on "Tests" it was pointed out that soft opaque white has a tendency to become green in the burned-out parts and where it is applied too thinly. Soft opaque greens, and particularly the opaque reds, vermilions, and oranges are also subject to color changes when burned-out. Remember that the outside edges of any bowl, plate, or panel receive heat from the kiln coils first and that the center of the piece becomes fused last. This is true whether the piece is suspended on prongs or resting on other firing devices such as pieces of Marinite or fire brick. Because the back of the kiln is hotter than the area near the door, to achieve the blackening or "burned-out" edges as a deliberate part of the design, you must remove the piece from the kiln, turn it, and replace it in the kiln several times.

FIGURE 63. "Sunflower," panel by the author; sgraffito and gold burn-out technique.

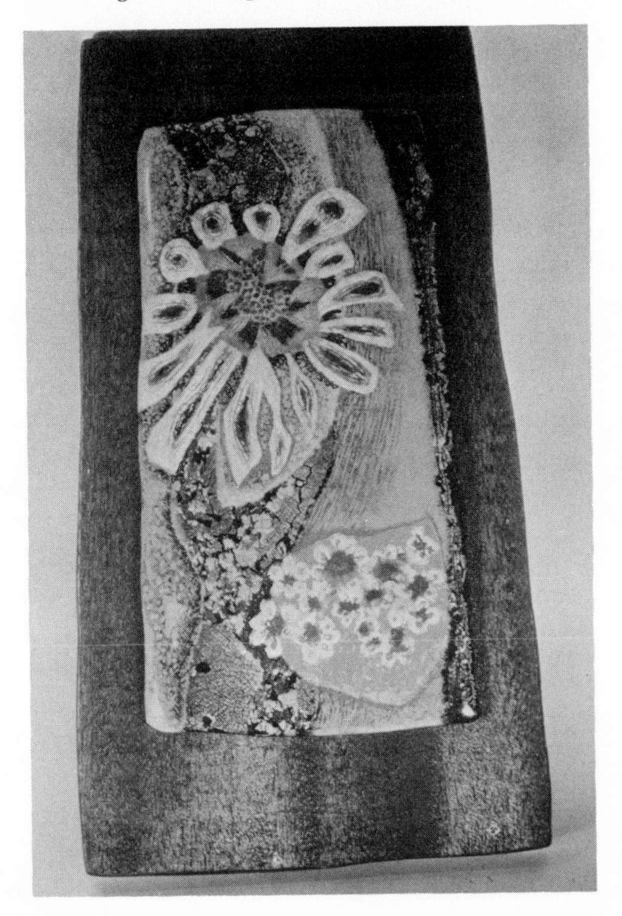

Intentional burn-out at the edges will often enhance the effect of an otherwise dull and pedestrian-looking piece. In spatula work, the soft reds (which burn-out) may be applied thinly and interspersed with harder colors (which will never burn-out) for the express purpose of creating interesting, "loose" textures.

Silver and gold foils applied in sheets or as paillons are burned-out sometimes by overfiring especially if the enamel base for the foil is a soft (low-fire) color. Strangely enough, the burned areas of the gold foil seem to show no ill effect when enamel is fired over them, but burned-out silver foil is another matter. In this case the enamel reveals the darkened areas. To utilize purposely the burned silver foil areas (that is, place it in a hot kiln, and allow sections to become black) seems quite legitimate as a means of searching for new textures.

Working with silver cloisonné wire made of fine silver has its hazards. Remember that the silver is in tiny strips and that it is in contact with the grains of enamel. Because of the delicacy of the wire and the heat to which it must be subjected (at least 1450° F.), melting of the silver wires is an all too common occurrence. The silver wires, when burned, actually become liquid and seem to flow under the enamel. Despite the sadness of such an accident, after further stoning and refiring, the result can be very interesting. All stiffness and hard edges, those characteristics that sometimes contribute to the rather mechanical look of cloisonné, are blended or fused into a surprisingly fascinating design.

The intentional use of burned-out enamels, metal foils, and cloisonné wire may be legitimately added to our list of experimental techniques.

6. *Fire Glaze*

There are many uses of fire glaze, or fire scale, which is the result of oxidization of copper when subjected to a heat above 1300° F. One way to use this technique is to clean a piece of copper and then place it in a kiln at about 1700° F. until the entire surface is deep pink. As the piece is withdrawn and subjected to cool air, sections of black fire glaze will fly off, leaving, at times, a most intriguing pattern of spots and textures. (If the copper is fired too high and too long, the entire surface will remain black, leaving no fire-glaze pattern). Now proceed to remove any loose particles of fire glaze by immersing in nitric acid (four parts water to one part acid) and scrubbing vigorously with coarse steel wool. The panel may now be considered as a background for further development with enamels. Either transparent or opaque enamels may be fired directly over fire-glazed areas.

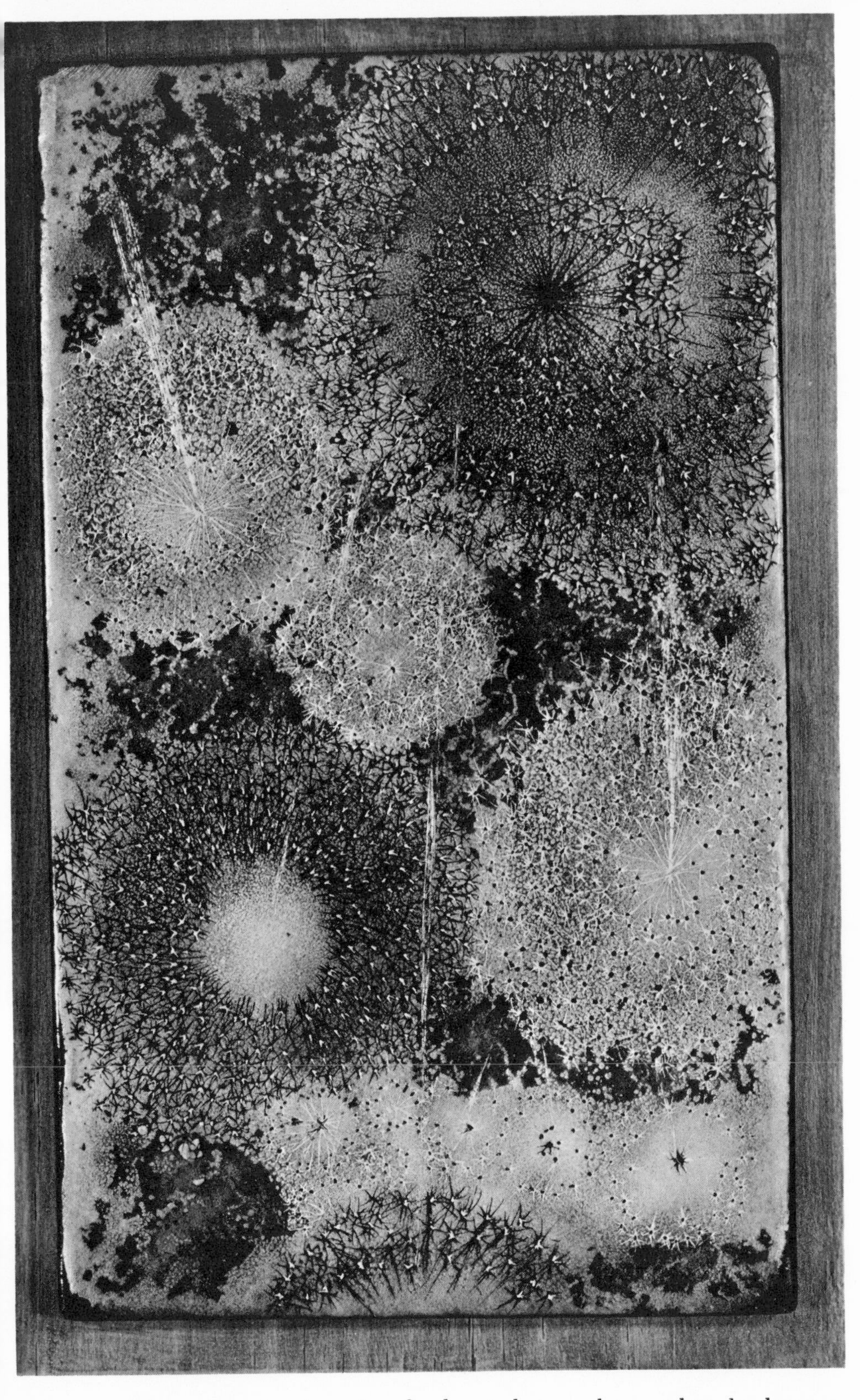

FIGURE 64. "Elegant Pest" by the author; fireglaze technique plus overglaze development.

Flux, when fired over fire glaze, becomes a reddish brown. Other transparent enamels are effected in color only slightly. Try the experiment of producing fire-glazed areas as described above and then polish the exposed copper areas to a high sheen. Proceed with a thin coat of well-washed transparent citron yellow (or try this same color over the fire-glazed line produced by sgraffitoing through liquid flux to the raw copper). Polish with steel wool before enameling.

Inevitably the studio bench and certainly the classroom bench are laden with fire glaze or fire scale particles, which should be constantly swept away to avoid the possibility of them falling accidentally onto an enameled surface. But because they so readily attach themselves to enamel and become imbedded in it, they constitute a definite potential for textured effects. Collect the fire scale from the enameling bench, and manipulate large or small shapes into an abstract pattern. Fire these onto the surface, placing them intentionally and proceed with subsequent firings of transparent enamels. Another method is to place the fire-glaze scraps in a screened jar and shake or dust these particles onto the unfired enameled surface. This produces not only some fascinating organic patterns but a rough, coarse quality, which relieves the monotonous glossy look of enameled surfaces. You could go further and actually draw with the fire-glaze scraps. (Such experimentation relieves the monotony of working continually with the known and traditional techniques.)

7. *Platinum Outline*

The occasion may arise where you will want to produce a very exacting, hard-edged interpretation. (A commission, for example, might demand such elements as lettering, insignias, heraldic plaques, trade marks, or some sort of commercially duplicated designs.) The concern here is not necessarily with unusual textured effects, accidents, or freedom of interpretation; the aim is to use the medium of enamel merely for its permanency or for its depth of color.

You need an accurate, well-defined working drawing, which must be rendered on tracing paper in the proper scale. The plaque, panel, or motif is cut to the proper shape and formed of 18 or 16 gauge copper. For such exacting work, it is well to keep the panel flat; however, a very slight capping (or turning-down) at the edge of the panel strengthens the piece and also gives it some quality. (Perfectly flat panels, unless recessed into wood or surrounded with some sort of frame, tend to look cheap and unfinished.) If it is possible, work on a white background for this kind of rendering. Use hard opaque white or ivory, applying several coats for a mechanically perfect surface. (The same would hold true if a black background is required.) Stoning and refiring will accomplish even greater perfection.

For the tracing it is well not to use an excessively greasy carbon paper, but one that will leave a fine line rather than a thick broad line. Red or white carbon paper is most satisfactory for white or black backgrounds. Experiment with the many varieties of carbon paper obtainable in stationery stores. (It stands to reason that graphite paper will not transfer to a highly glazed surface such as enamel.) Fasten the tracing at the top of the enamel plaque with Scotch tape, insert the carbon paper from below, and trace with the steel pointer or jeweler's stylus. The next step is to retrace the carbon line with liquid platinum or palladium luster. For this a Hunt's globe bowl-pointed 513 pen or a 303 Gilott pen point is best. Allow the palladium to dry thoroughly by forced drying or by letting it remain in a normally heated studio over night. After firing the lines at 1500° F. for about two minutes, the carbon should disappear entirely, leaving a clearly defined blackish metallic line on the white or black background.

Develop the colors by a dusting method. First wet the surface of the

FIGURE 65. Platinum line technique; double shape from one piece of copper, by Josephine McCorckle.

enameled panel, and by the use of small bottles with 80 or 100 mesh screening inserted into their caps, proceed to build up the required tones. Shade the necessary areas by dusting before firing. Soft edges are the result of dusting, but to create hard edges wipe away with a moist brush. Work as many colored areas as possible at one time, fire at a medium temperature, and proceed with the remaining parts of the design. Subsequent transparent colors of various tones are used to enrich the original scheme. By using the atomizer with sudden jerks close to the dry dusted areas a spotted texture can be produced. It is obligatory when employing the above technique that the dusting should be very fine (100 mesh screening preferably) and applied in extremely thin layers or coats.

Silver or gold paillons may also be incorporated with this method. They would, of course, have been fused to the opaque base color before the dusting of transparent colors.

8. *Sponging*

Sponging (which may be called a device rather than an actual technique) offers endless possibilities and, in contrast to outlining, is a very loose or flexible manner of working.

It is difficult to dissociate painting from enameling, considering the

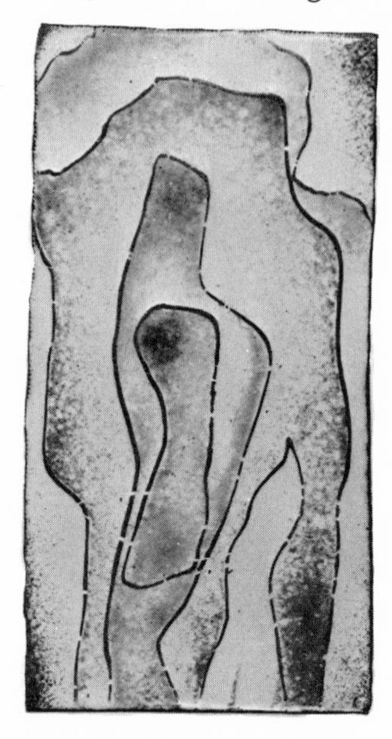

FIGURE 66. Platinum line.

multitude of contemporary mannerisms they have in common. Indeed, the narrow line between a painting and a decorative panel is never too clearly defined. Here you will build up under-painting and over-painting glazing in the same way you would proceed with oil paints.

Prepare the panel or plaque with whatever ground you wish to work on: opaque white, transparent flux, colored transparent, or colored opaque. This ground may be unevenly applied so that when it is fired (perhaps at a high temperature of 1700° F.) some areas become burned-out, suggesting patterns with considerable potential for further compositional exploration. With a sharp razor blade cut directly through a sponge, thus creating flat surfaces. Some sponges will be found that have excessive porosity, some with large holes, and some with small holes. (Make a collection of various kinds of sponges, some natural and some synthetic. Include those sponges that are not cut to make a flat surface.) Pour out a small amount of squeegee oil in a flat saucer or dish, dip the sponge into it, and stamp or print the oil directly

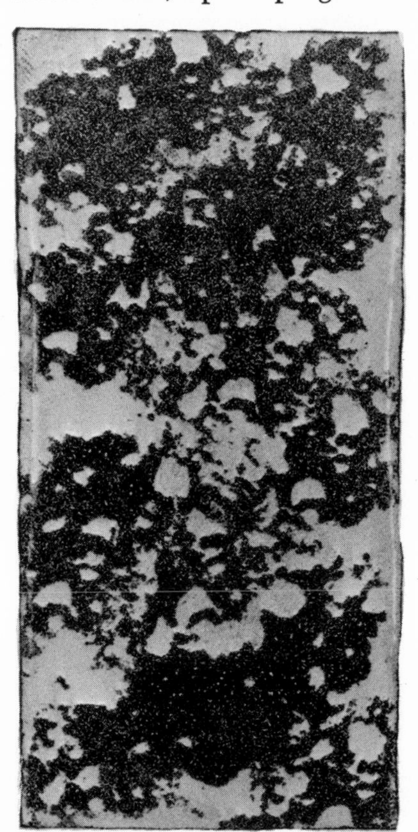

FIGURE 67. Sponging.

onto the copper. Using an 80 mesh screened basket, dust the entire surface. Dump or blow off that which does not adhere and the resultant pattern should prove to be surprisingly interesting. Fire this, clean with acid, and proceed with subsequent spongings and firings.

Sponging is usually done by firing first a base coat of either transparent flux or hard opaque white. From there you may proceed with numerous coats of sponged surfaces, developing the design as you go along, with the intention of relating or organizing each added texture to that which has already been established. A thick solution of gum tragacanth, made by dissolving the powdered gum first in wood alcohol and then slowly adding water, is equally as successful as the squeegee oil for the sponging.

Another use of the sponging technique is to apply it over a previously hard-edged spatula design. The added sponged textures may be incorporated into the design, building up fascinating textures. It is wise to make use of the opaque colors first, following these with the transparent enamels to give depth and interest to the various areas.

9. *Stenciling*

Stencils in enameling have their place, although one usually thinks of stenciled renderings as stiff and characterless. But stenciling is justified in certain instances—such as large murals, or commissions requiring dozens of duplications.

For the individual piece, a method of stenciling is as follows: A clean-cut and accurate working drawing is used as a pattern for the stencil segments. Fire the base coat, whether it is transparent flux or any other

FIGURE 68. Hand-hammered tray by the author showing simple stencil technique.

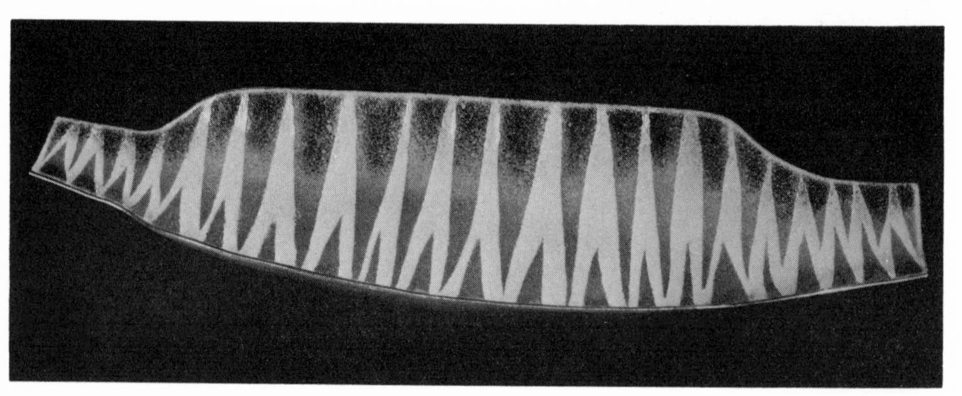

predetermined color. Cut the stencil patterns from ordinary paper toweling by tracing the working drawings. Even better than paper toweling is a non-woven cellulose material called Pellon that does not expand when wet. Place the stenciled shapes upon a wetted surface of fired enamel and dampen with the atomizer (the stencil paper should not be too wet—nor too dry). Dust the enamel through a 100 mesh screen in either solid or shaded areas, and then withdraw the paper stencils, using a pair of pointed tweezers. By ingenious manipulation of stenciled areas and even stenciled lines, a commission of one hundred trays in exact duplication can be accomplished. Perhaps two or three firings might be employed, but the craftsman is able to expedite this kind of job in a much shorter time than if he were to resort to the spatula method (wet inlay).

Because stenciling in any medium is rather hard and inartistic-looking, I would like to present a variation that might be of help. Instead of laying the stencil shapes—whether paper or thin metal—down onto the surface of the fired enamel, try placing small blocks of wood (or perhaps jar covers) outside of the piece to be stenciled. These should be about three quarters of an inch high. Cut the stencils

FIGURE 69. Interesting edges may be made by stenciling with torn, rather than cut, paper; gold luster crackle.

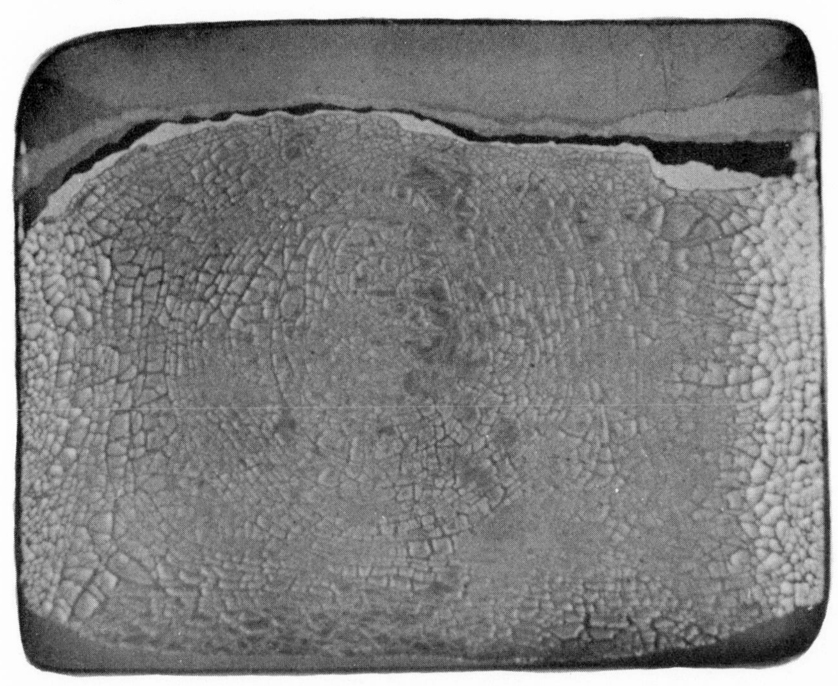

of thin cardboard or construction paper and place them so that they straddle the blocks. When the dry enamel is sifted over these raised stencils, the areas are softened or diffused at the edges. By dusting and blending in this manner all semblance of the hard, mechanical look so objectionable in stenciled designs is avoided.

10. *Lumps, Threads, Scrolling, Separation Enamels*

I have purposely grouped the use of lumps, threads, scrolling, and separation enameling as techniques. During the recent phase of enameling as a hobbyist's craft, these four contributed much toward arousing the art critic's disfavor. More than any others, these particular techniques, when in the hands of the rank amateur, transformed enameling into a kind of popular-activity contest, made the craft seem too easy, and produced superficial results. (It is my belief that some of this phase has passed, and though some bad enamels may always be with us, still, there are those who wish to improve their taste and who wish to look at enameling as a more serious art vehicle.)

FIGURE 70. Stenciling; raise the template with jar covers to create a soft stenciled edge.

I would like to defend these techniques to a degree. My criticism of them above, as making enameling seem too easy when in the hands of the rank amateur should be qualified to say, "the insensitive amateur," as it is not my intention to underestimate the untutored or the "unprofessional" artist. Indeed, genuine feeling sometimes comes from untrained artists. Any technique, regardless of its easiness or difficulty, is not to be thrown aside simply because it may have been misused.

Lumps or chunks of unground enamel may be placed on the fired enameled ground and attached by firing at low or high temperatures. It is wise to use a gum tragacanth solution to hold these chunks of various sizes in place when firing. When fired high, they sink or melt down, giving areas of brilliant jewel-like colors. With lower firing, the chunks tend to have a raised effect and resemble cabochons, which were often seen in the late French Limoges period.

Sometimes by accident, the enamel, while in a molten state, becomes

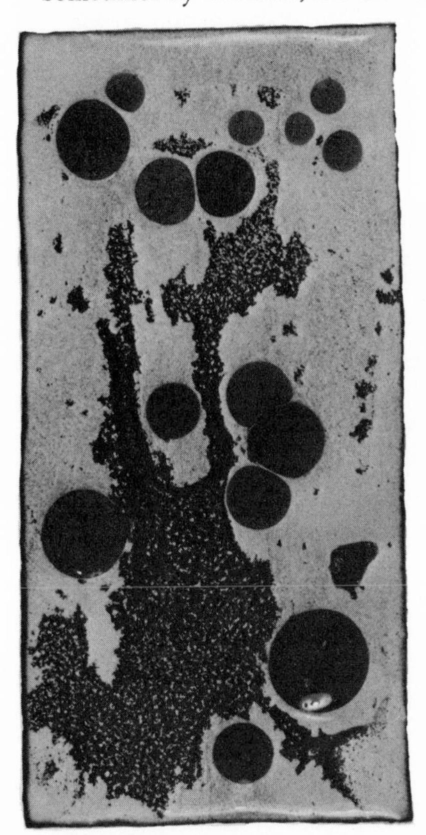

FIGURE 71. Lumps.

attached to the metal trivet or to the tongs. Small threads occur as the enameled piece and the trivet are pulled apart. These delicate threads of any colored transparent or opaque enamel are readily fired onto a previously ground-coated surface.

Both the threads and the chunks can simply be dropped by handfuls onto any plate or plaque and fired. Unfortunately, there are large quantities of such enamels produced in a thoughtless manner, but if planning and a sense of design accompany the procedure, there is no reason why such a technique should not be employed. Enamel threads can be bought; sizes, shapes, and lengths can be selected and the design can be the result of serious compositional planning.

Scrolling is another easy-to-do technique. It consists of making use of a long-handled steel tool with a pointed hook on its end. This hook is thrust into the kiln, and by scrolling or swirling the enamel while in its molten state a kind of marbleized quality results. I must admit that, although this technique rapidly became a hobbyist's delight, there are a few enamelists in the country who, with great skill in application and firing and by careful selection of colors, have produced some extraordinarily handsome effects with scrolling. However, this, like any of the "easy" techniques, is not foolproof.

To develop the separation enamel, you must first fire a rather heavy coating of a substantial counter-enamel. A light application of any soft opaque enamel will not be sufficient for the counter-enamel because

FIGURE 72. Threads.

eventually the piece must be raised to a high temperature in the kiln. On the face of the bowl or plate (it is wise to try the separation enamel first on a deep bowl) fire an even and fairly heavy coat of soft fusing flux. Next, with the spatula apply solid areas and spots of almost any enamel—near together, far apart, or touching, it doesn't matter. Fire these at the normal temperature of 1500° F. Now, with the wire mesh basket, dust on a layer of any dark transparent enamel entirely covering the other layers. Fire this at the same temperature. Finally, with the liquid separation enamel, which can be purchased at any

FIGURE 73. "Tensions," panel by the author showing threads technique. (*The Cleveland Museum of Art, lent by the artist for the thirty-fourth May Show*)

enamel supply company, thinly paint on a linear design using a medium-sized water-color brush. This design may be quite conventional in nature; that is, stripes, bands, spots, etc. arranged formally around the bowl. These lines should be brought up *near the edge* of the deep shape for, as you give the piece a final firing of at least 1700° F. for two or three minutes, the separation enamel will cause the undercoat to run down, completely distorting the conventional application of lines. Although this procedure may appear to be nothing other than therapy for rainy afternoons, in its defense may I say that certain artists have, through considerable research and endless tabulation of experimental tests, so perfected this technique that their end results are worthy of any exhibition.

FIGURE 74. Separation.

11. *Basse-taille and Repoussé*

Basse-taille enameling is one of the older techniques, excellent examples of which can be seen in many museums. The most famous example is the Royal Gold Cup, made in Paris as early as the middle of the fourteenth century and now at the British Museum.

For the basse-taille process use metal of 22 gauge. After cutting the shape and cleaning the surface, place the metal on a block of soft wood or the sandbag, or a combination of both as the work progresses. Use the chasing tools, or similar tools that are easily improvised, in conjunction with the chasing hammer to countersink lines, shapes, edges, and textures. As you work from both the back and front sides, annealing helps to keep the metal malleable. Tooled surfaces are often developed by a variety of steel punches, and engraving the surface gives further enrichment. The tool is held under the hand and guided with the thumb. Keep the surface clean by increasing the acid solution to about two parts water and one part acid or even stronger. Polish with very fine steel wool and when possible do the engraving last in order to produce a brilliant sparkling quality. Sometimes flat dulled chisels or corners of steel blocks and many other odd shapes can be used creatively.

The enamel is applied to both back and front of the piece and slowly built up high enough to be stoned to one level with the carborundum

FIGURE 75. Basse-taille hammered tray by June Schwarz; acid etching beneath the surface of the enamel. (*American Craftsmen's Council, New York*)

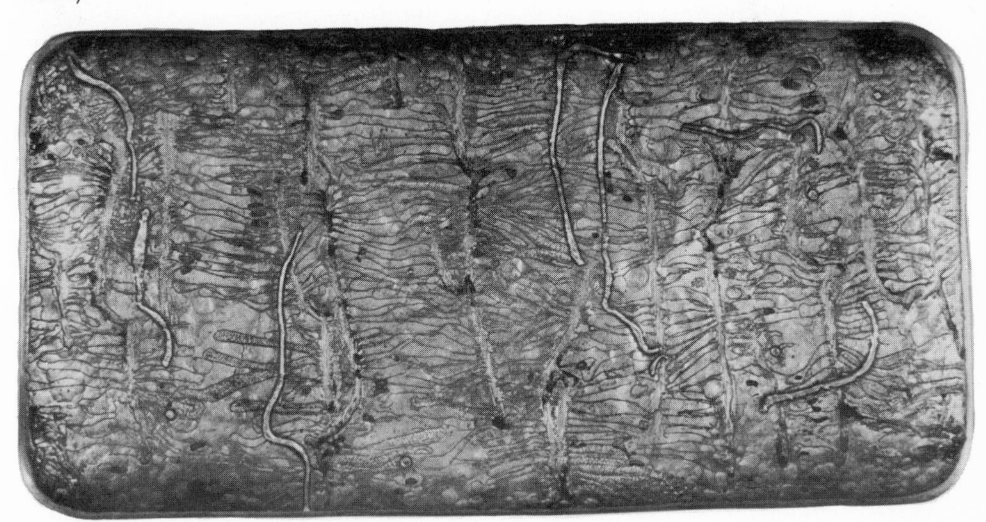

stone. Be careful not to stone through on the high places. Also, because of the thin gauge metal used it is advisable to strengthen the piece by firing a piece of any iron or steel wire mesh to the back, fusing it into the counter enamel. The enamels must extend over the entire surface giving a richer color where it is deepest. As only the most transparent colors should be used, the effect achieved may even be too dazzling. To avoid this, the surface can be rubbed with a fine grained Scotch hone, and brought to a softer finish by buffing with tripoli or extra fine pumice and water. This constitutes the true basse-taille type of enameling.

A variation of basse-taille, and one closely related to it, is a technique known as repoussé enameling. Here, the same procedure of modeling the metal by chasing and engraving is followed, but the enamel is not necessarily brought to one level and stoned. The piece shows its topographical or sculptured surface.[1] Certain areas of the

[1] Helen Worrall, "Enamel on Copper Repoussé," *Ceramics Monthly*, March 1964, pp. 12–14.

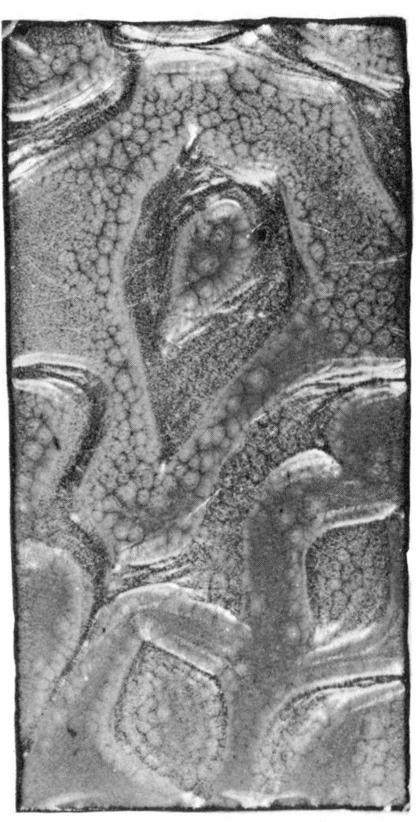

FIGURE 76. Repoussé.

metal may be exposed such as a margin or surrounding border. In this case the chasing tool is first used to sink the whole area of the motif after which modeling is done within the desired area. By enameling over the uneven surface you are actually incorporating what is known as "encrusted enameling." Figures in the round were often seen with *encrusté* enameling, probably the best-known example of which is the Benvenuto Cellini salt cellar. Adaptations were done in Vienna and occasionally in this country in the form of small decorative animals or more abstract twisted motifs.

12. *Grisaille*

The word "grisaille" comes from the French word "gris" meaning "gray"; hence the enamels of this technique are dominantly neutral in color. The technique may be observed in many circular plaques and

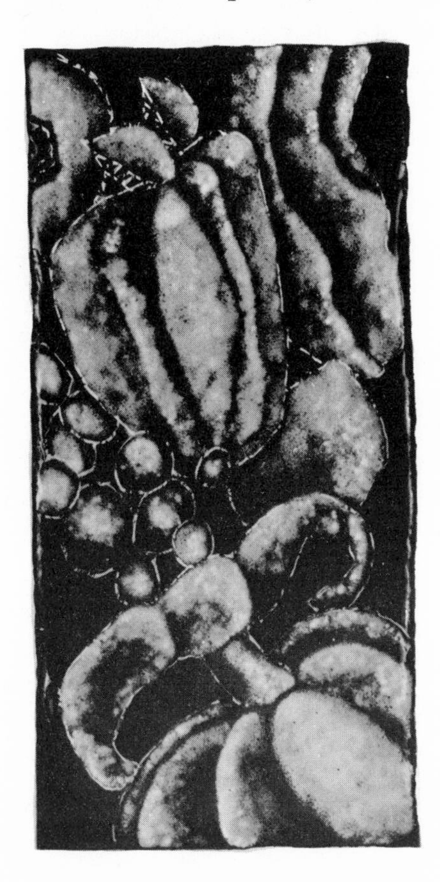

FIGURE 77. Grisaille.

illustrative panels dating around the sixteenth century and coming from the Limoges or nearby districts.

Because the process involves a kind of underpainting, or build-up by values, for depth and structure, amateurs or those with little skill in drawing would be wise not to try it. Actually, this is one classical technique that has been explored very little by the contemporary enamelist, although, in my estimation, it has infinite possibilities.

Start by applying a thin coat of opaque black, or perhaps some other dark color. Fire this, and stone the surface level for a smooth working area. Fire again after stoning. Then draw the composition with a pen or tiny brush, using one of the metallic lusters or perhaps a bit of overglaze; fire again in order to proceed with the design.

FIGURE 78. Two scenes from the "Aeneid" by Jean Pénicaud, French (1520–1588); grisaille technique. (*The Cleveland Museum of Art, gift of Robert A. Weaver*)

Use your hardest opaque white. First sift it through the finest mesh, or use an improvised screen made from a nylon stocking. Place this finely ground opaque white in a small agate mortar and using the pestle grind it most assiduously with water. (Washing is not necessary with opaque white.) Now, after draining, place the enamel on a surface similar to a ceramic tile, add several drops of oil of sassafras (or oil of lavender) and one drop of squeegee oil, and with an artist's palette knife grind it further. The final result should be a pastelike substance not unlike tempera paint. You are now ready to start the underpainting or modeling of the surface. This is not done in one firing. Because a three dimensional effect is desired, continue to "build-up" the whites, progressing from thin coats, which give the darker grays, to heavier coats, which produce almost dead white for the highlights and nearer planes.

By firing very delicate layers of finely ground and washed transparent colors, or perhaps by adding a few gold or silver paillons and details with black overglaze and metallic lusters, you can achieve a comprehensive dimensional drawing. Although in the past this technique was used chiefly for representational work, it is one that might reasonably be used by the present-day abstractionist.

13. *En résille*

En résille is another classical technique not too often seen. Actually *résille* means *hairnet,* and, in fact, the delicacy and fragility of this rare kind of enameling are not unlike the delicacy and fragility of a hairnet. Instead of being fired on metal, some of the en résille enamels are made on a polished piece of crystal. Close viewing reveals that the fine tracery is not the result of bent gold wires, as with cloisonné, but is produced by incised lines in the crystal, which are subsequently filled with fine gold.

There are two pieces of en résille enamel in the Gallerie d'Apollon in the Louvre, some in the South Kensington Museum, a miniature case in the J. Pierpont Morgan collection, and a few examples in private collections. This type of ornamentation, delightful in effect and of a marked delicacy, belongs properly to the French goldsmiths of the

FIGURE 79. Detail of "Bowl of Gold" by Margret Craver, 1961; attached motif in en résille enamel.

second half of the sixteenth century, but no one knows who actually invented it.

The en résille technique was used for jewelry, for mirror or miniature cases, and for objects of adornment. As Dr. William Milliken, director emeritus of the Cleveland Museum of Art, once pointed out, it was indeed a curious technical process. The desired design of arabesques and scrolls was engraved upon the surface of a plaque of glass paste or rock crystal to a depth of perhaps a half a millimeter. This incised pattern was then lined with gold and the tiny compartments so formed were filled with opaque or translucent (transparent) enamels of an extreme fusibility (very soft, low fired colors). The result, after careful firing and polishing, was a beautiful all-over design almost like a kind of incrustation upon the crystal surface. Obviously, because of the excessive fragility of the materials used, the process was one of great delicacy. Margret Craver, who is one of the few craftsmen in this country exploring the en résille technique, says this about it: "After studying the few pieces extant of this sixteenth-century technique, I began to think of it as a reverse process—i.e., the gold floating on top of the gold."

14. *Individual Techniques*

a. ENTIRE FOIL SURFACE. In order to avoid the hard-edged, cut-out look of paillon shapes, the following individual manner of working is suggested: First fire a thin layer of transparent yellow over the entire surface. Next place full sheets of either the gold or silver foil between tissue paper folded at the bottom for ease of handling. With the wallpaperer's sharp toothed roller pierce tiny holes throughout the foil sheet. This is a much easier method than punching the holes individually with a pointer or jeweler's scribe. Wet the surface of the enamel with thick gum tragacanth and carefully lay the sheet of foil down. (The metal foil may overlap in places.) After removing all moisture with absorbent cotton, fire at about 1500° F. Firing the foil at too high a temperature or too long will cause the large sheets to withdraw from the edges of the panel, so be careful. Next trace the design onto the fired foil using red or white carbon paper, and proceed to wet inlay all motifs with transparent enamels and fill in all background areas with an opaque enamel, bringing the colors together.

The first firing may be disappointing as the enamel will crawl away from the foil in small spots. The foil, being pure metal, will not oxidize so you may re-enamel the identical design, both with the transparent colors and the opaque background. After the final stoning an unusual effect is achieved, that of soft-edged transparent shapes and a kind of semi-transparent background as the silver foil sparkles through even the opaque colors.

b. FLOWING COLORS ON HARD WHITE. An interesting effect that approximates a water-color wet wash is obtained in the following manner: Apply an undercoat of your hardest opaque white over the entire surface. (There is a special hard white enamel for this purpose.) Grind and wash an interesting palette of transparent enamels and proceed to move them freely about with the spatula, pushing them into rather loose and amorphous shapes with no attention given to hard, exact edges. Flood the whole plate with water from the atomizer until it is much wetter than is normally used. Then take a tool, such as a small hammer or heavy brush handle, and tap from beneath the plate,

FIGURE 80. "Garden Symbols on Gold" by the author; finished tray, with water-color sketch below.

FIGURE 81. Step 1. Preparing panel. (*Figures 81–93, courtesy* Ceramics Monthly Magazine, *Feb. 1967*)

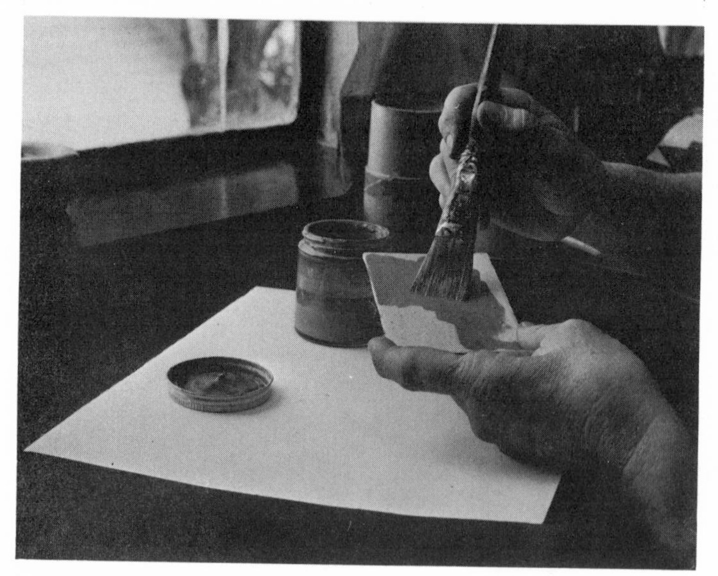

FIGURE 82. Step 2. Applying counter-enamel.

FIGURE 83. Step 3. Applying scale-off to fired counter-enamel.

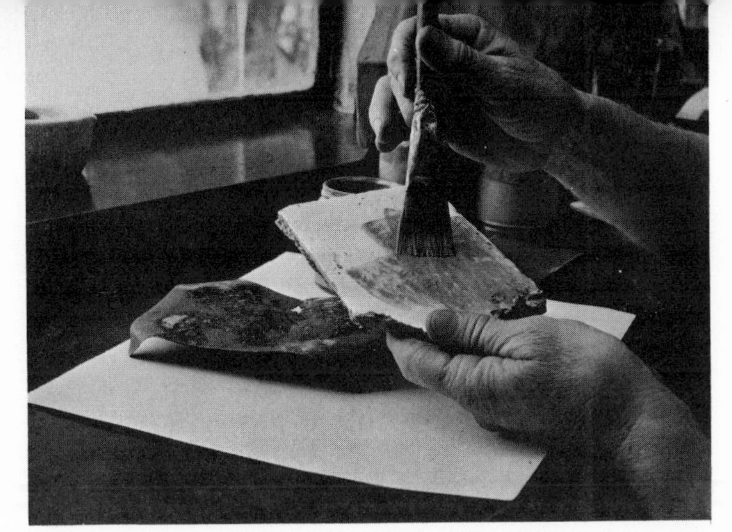

FIGURE 84. Step 4. Applying scale-off to marinite firing block.

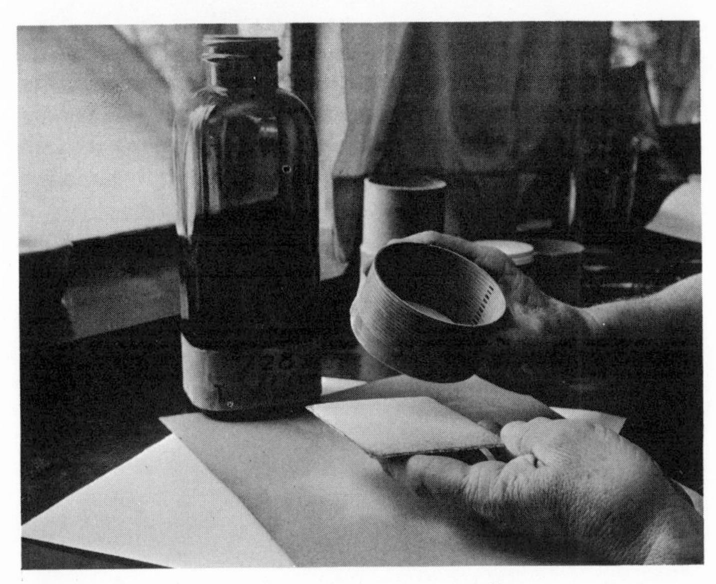

FIGURE 85. Step 5. Dusting undercoat of transparent yellow.

FIGURE 86. Step 6. Piercing foil with roller.

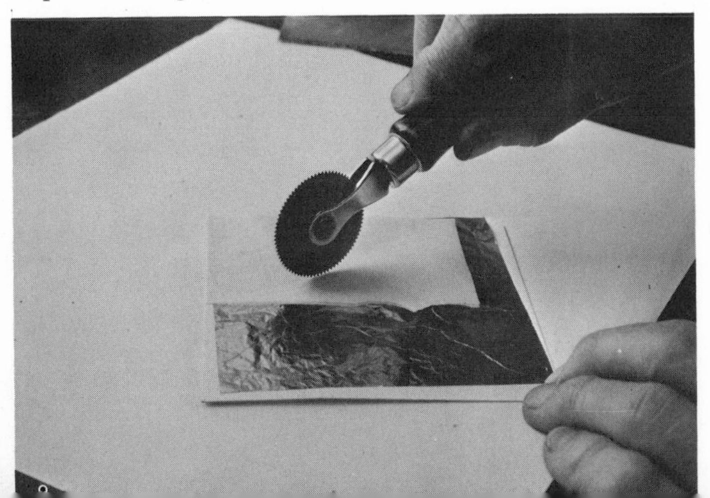

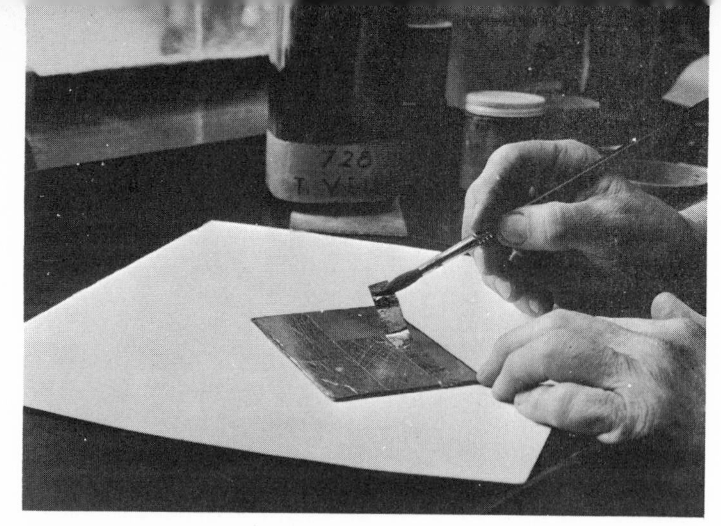

FIGURE 87. Step 7. Applying metal foil over entire surface.

FIGURE 88. Step 8. Tracing design.

FIGURE 89. Step 9. Grinding each color with mortar and pestle.

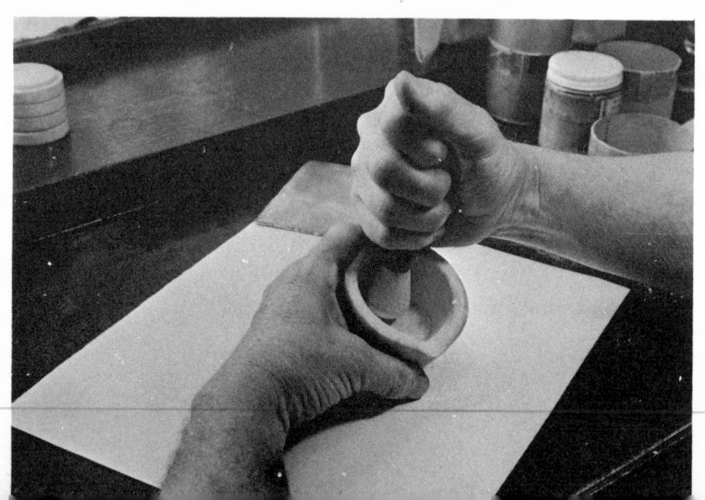

FIGURE 90. Step 10. Applying first coat of enamel by spatula method.

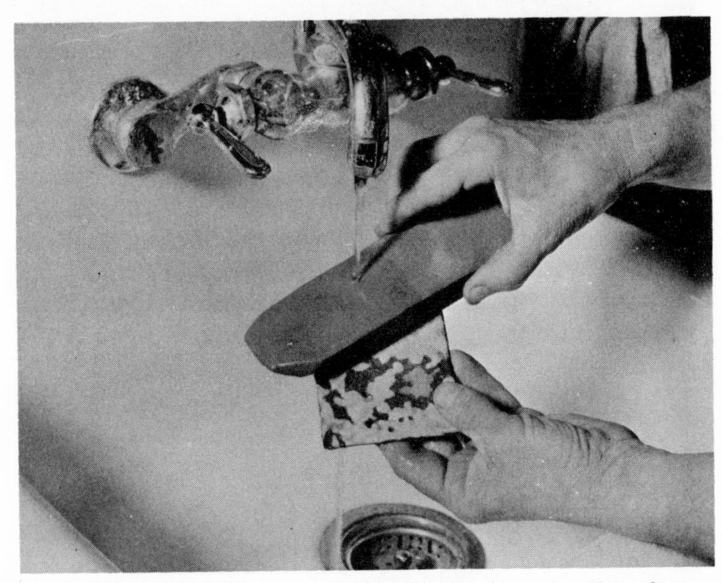

FIGURE 91. Step 11. Stoning under water in preparation for second application of enamel.

FIGURE 92. Step 12. Applying overglaze line after second application of enamel.

FIGURE 93. Step 13. Final firing of plaque.

controlling the flow of the wet colors. After slightly drying, you may, if you like, sgraffito back to the white ground in some places. Fire the panel, stone, add other transparent colors in the same wet saturation manner; fire, stone again, etc. The effect is a relief from the more traditional spatula work.

c. MULTIPLE FOIL LAYERS. Added depth of color may be achieved by adding more than one layer of foil and enamel. Transparent Ruby Red is a good suggestion for this individual technique. First fire transparent red directly onto copper, which will produce a rather disappointingly dark maroon color with little life or vibration. Next, apply a patina of gold foil paillons about 1/8″ square, which are easily made from the sheet of foil by cutting in two directions. These tabs are picked up with a moist brush and applied, one by one, creating a close-patterned surface, allowing the dark color to show through occasionally. Fire. Apply an even coat of well-washed ruby red, evaporate moisture, and fire again. Next, place a second layer of gold paillons onto the surface, allowing the two tones of red to be visible in places. These paillons are again covered with transparent ruby red and finally fired, stoned, and refired. Unusual depth of color will result as the various levels and tones become visible.

d. WISTERIA LINE. Richness and transparency are the result when certain fugitive opaque colors are purposely burned out. For example,

FIGURE 95. "Ruby Urn" by the author; an example of the use of two layers of gold paillons to achieve depth of color. (*The H. O. Mierke Collection*)

FIGURE 94. Detail of panel "August 15" by the author; wet transparent colors over hard opaque white.

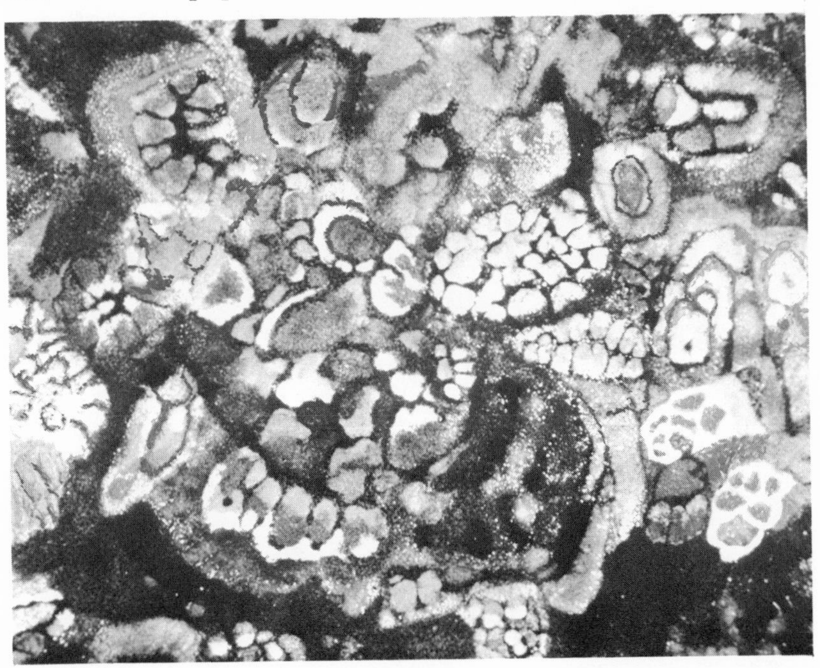

try firing a coat of soft flux over either a plain surface of copper or an engraved or vibrated surface. If desired, metal foil paillons may be added at this time. (Before the foils are attached, they may be broken or scratched with the pointer for greater textural interest.) Next fire a thin application of light transparent colors into the design. Now, a particular transparent color—namely wisteria, a soft brownish purple—is dusted thinly over the whole surface and also fired at normal temperature (1500° F.). Lastly a very light coat of one of the softest opaque colors (such as opaque gray, turquoise, or chartreuse) should be sifted over the whole panel, and with a pointer or sharpened brush handle a linear pattern should be sgraffitoed back to the wisteria color. Place the enamel in a kiln with a high temperature (1600° F. or more) and continue firing until all of the opaque enamel has turned to transparency, leaving a fascinating tracery of the wisteria color. Be-

FIGURE 96. "Women," panel by Fern Cole, 1959; sgraffito with burn-out (wisteria line technique). (*Museum of Contemporary Crafts, New York*)

cause the wisteria enamel is "stable" (does not burn out) and the soft opaque enamel is "fugitive" (loses its opacity and becomes transparent with high firing) this technique lends itself to most fascinating experimentation.

e. USE OF STRINGS. A technique worthy of further exploration is that of using strings to stencil off colors. Certainly, in the hands of an artist whose interest is that of a painter, this process should hold some fascination. Generally speaking, much of enameling as we know it today *is* painting. To me, it seems unfair to speak derogatively of craftsmanship, but respectably of painting. In the fabrication of Op paintings, craftsmanship plays a very prominent part, and in many of the recently developed techniques in enameling the result is a veritable "painting in glass."

The use of strings has many possibilities. You may start with a black background or a light background, and play with a variety of sizes of strings, ropes, threads, yarns, and pliable wires. Place these upon the surface of the fired background. (Dust opaque enamel if the background is black or transparents if you are working on a light ground.)

FIGURE 97. "Owl" by the author; the use of strings or threads as stencils to create lines.

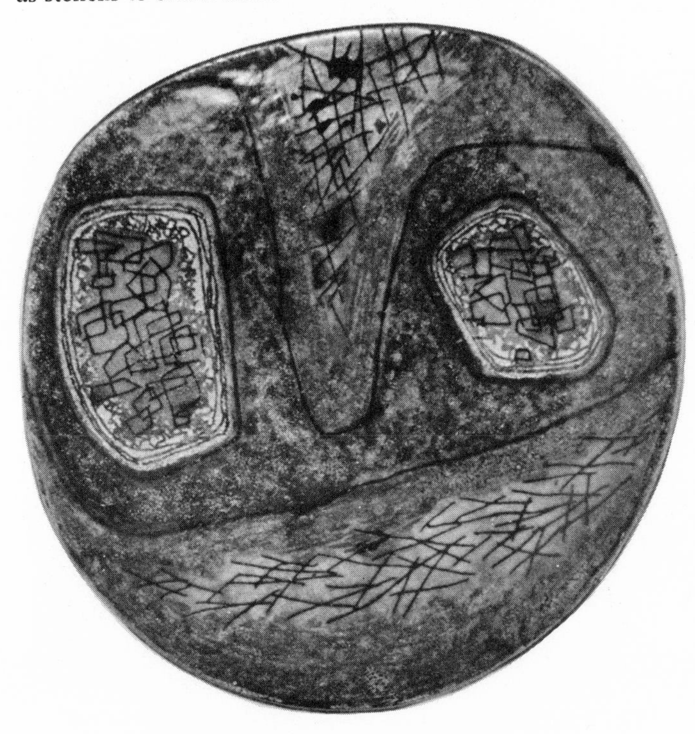

Continue the process of blending colors, firing, adding other strings, firing again, etc., building up a design that can have considerable spatial quality.

f. FIRING WITH BLOWTORCH. The enameling kiln or muffled furnace is not necessarily the only way of fusing the ground enamel frits. For small experimental tests or even for composite sections of a miniature box or plaque, you may improvise a handmade muffled furnace from two tin cans, a piece of wire mesh, and an ordinary gas blowtorch. This same technique may be carried much further; in fact, large screens and murals have been made without the use of a kiln. By the ingenious method of suspending sheets of copper over an open stand or grid, and by moving a series of gas flames or Burnzomatic torches under the surface, the enamel is fused. Areas of exposed copper may be left, or scales of oxidized copper incorporated into the pattern. Sgraffito, dusting, burned-in edges, and partially fired areas are all part of this

FIGURE 98. Enameled panel for screen by Paul Hultberg; enameling on a large scale. (*American Craftsmen's Council, New York*)

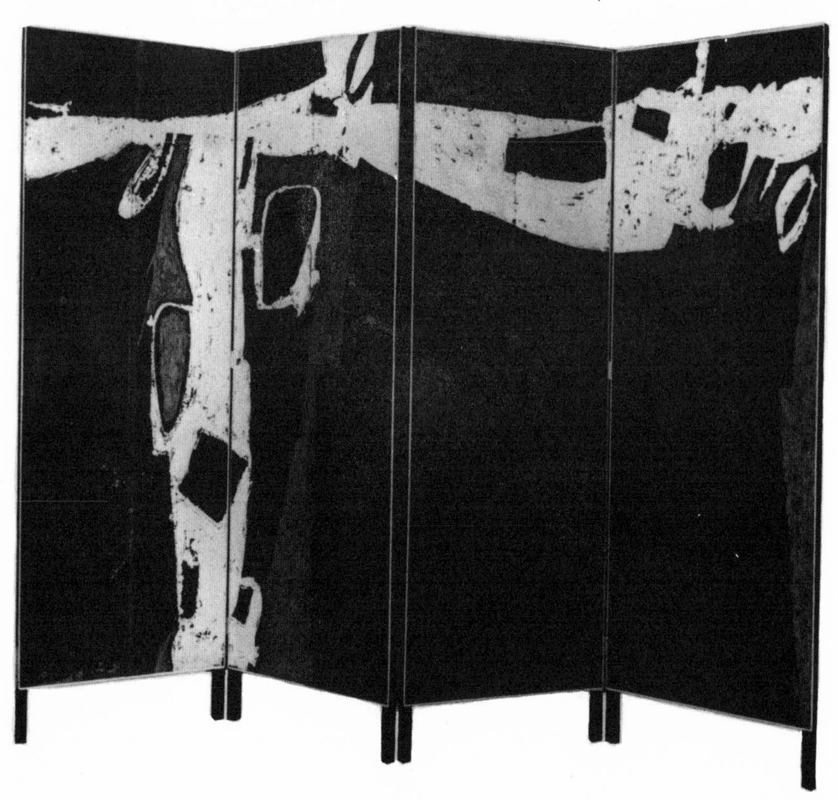

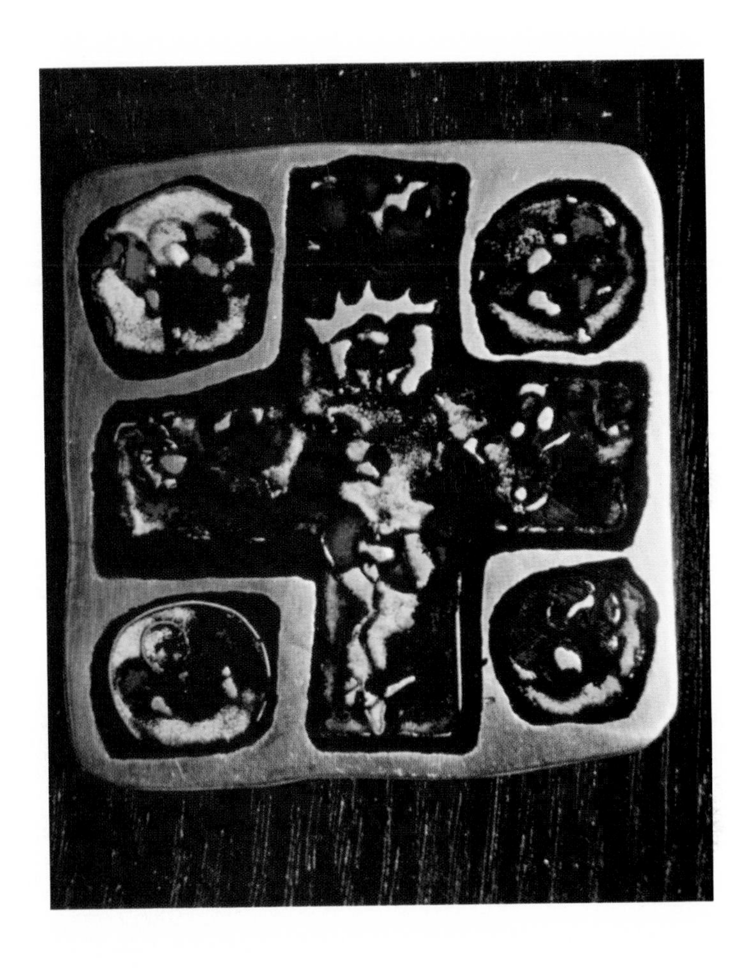

"ALPHA AND OMEGA" by the author; gold plated "burn-out" cloisonne. (Courtesy Joseph McCullough)

inventive and excitingly contemporary use of enamel. The outstanding achievement is, of course, that areas larger than any which might be subjected to the average kiln are within the range of the artist.[2]

These, then, constitute a few of the experimental techniques. There are many others that have been tried and also many others that will be forthcoming. Enameling is not a static art expression; it contains vast opportunities for further growth. For the contemporary artist (and let us terminate the endless argument concerning the difference between an artist and a craftsman) here is a vehicle, the possibilities and the ramifications of which are far from being exhausted. It is one medium that offers color, design, draughtsmanship, fusibility, use, depth of tone, scale, translucency, tactility, and a justification for skill or craftsmanship.

[2] "Paul Hultberg: the Enamel as Mural," by M. C. Richards, *Craft Horizons*, March/April 1960, pp. 29–32.

CHAPTER 5

ANOTHER LOOK AT CLOISONNÉ

1. *Wires*

Because the cloisonné technique has been known for centuries, much has been written about it. Here, I should like to consolidate the data and also present some new adaptations of cloisonné to contemporary design.

Originally, cloisonné enameling—which, incidentally, prefaced painted enamels (Limoges technique)—was a device invented to separate or define color areas. Because the enamel in its fused state became liquid or flowing, it was thought necessary to control the design by enclosing each color within a fenced-in area or cloison. Most of the early Byzantine, Chinese, and French cloisonné enamels show that the wires were rectangular ribbons set on edge and soldered (or in the case of pure gold fused in some mysterious way) to the base sheet of metal. This technique will be discussed later (page 143) in the section on gold cloisonné, but for practical purposes, most modern cloisonné work is done by simply fusing the wires to an enameled base, thereby eliminating the soldering process.

Wires may vary in many ways. For general work, an average size of ribbon wire might be 30 gauge wide by 18 gauge high or .010 inch by .040 inch. However, such generality hardly satisfies the creative craftsman. Wire as fine as 34 gauge by 22 gauge, or .006 inches wide by .025 inches high is often employed, and much heavier rectangular wire (20 gauge wide by 18 gauge high) is effective on larger pieces. It is not a good plan to purchase cloisonné wire higher than .040 inches (18 gauge) as the layer of enamel should not exceed this depth (thickness). We are speaking now of the type of cloisonné where the wires and enamel will be stoned to one level.

Round wire has some advantages and also some disadvantages. The advantage of using round wire is that, in bending the shapes, you need not be concerned about its standing upright, as in the case of ribbon or rectangular wire. (You don't have to turn a corner in order to make the wire stand up.) With round wire the design can be composed of single straight lines, which obviously must be avoided with ribbon wire. The difficulty with round wire, however, arises in the final stoning. Recognizing the fact that you are stoning down on a circular shape (the cross section of round wire) you must keep the stoning at the same level; otherwise, the line of wire becomes varied, or smaller, as it is stoned below the diameter of the wire.

In a contemporary use of cloisonné, a variation of the older technique is to leave the wires high or projecting from the surface of the enamel, with little or no stoning at all. In this case, the 18 gauge high ribbon wire is used with only one layer of enamel. When working on a fine silver base this method produces especially brilliant effects.

Square wire can be purchased in any dimension, or it can easily be fabricated by drawing the round wire through a draw plate. Square wire gives the effect of a broader or bolder accent line. It may be used with the ribbon wire, but remember that the height of the square wire should be similar to the height of the ribbon wire when combined in

FIGURE 99. Cutting round cloisonné wire, with separate piles for pieces of various lengths.

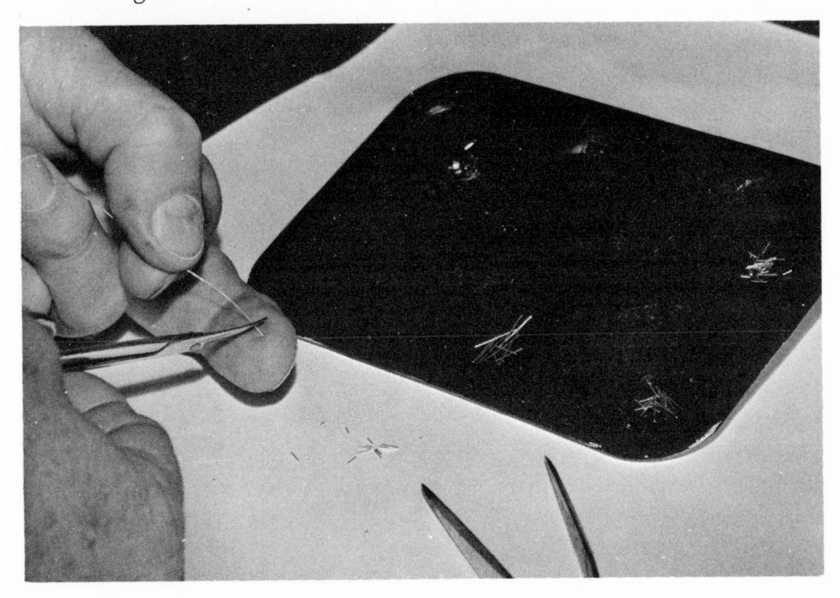

the same design to allow both types of wires to reach the same plane or level when finally stoned and polished.

Here is a way to escape the smooth flowing line so constantly seen in cloisonné enameling, and to approximate a more free, broken, sketchy line: Use 26 gauge fine silver round wire. Straighten the wire with the thumb nail or by pulling it over the edge of the bench and snip it with the manicure scissors (the metal cutter gives a better result). The wire should be cut in lengths ranging from $1/32''$ to $1/8''$ according to the motif to be rendered. Keep the various lengths of wire in separate piles or dishes. Moisten the surface of the enamel with very thick gum tragacanth (as thick as heavy cream) and pick up the tiny pieces of wire with sharp tweezers, laying them individually along the traced line. For the sharper curves you will need the tiny segments; for the straighter lines, longer sections of wire. This type of broken line adds "another look" to cloisonné and seems to take less time than bending the wires.

FIGURE 100. "Christ and the Evangelist," by Charles Barbley Jeffery, 1953; votive shrine with silver cloisonné wires raised above the surface of the enamel. (*Museum of Contemporary Crafts, New York*)

Although copper wire is not generally used in cloisonné, it should not be entirely disregarded. There are many color schemes involving reds, black, white, and browns, which would be more attractive with copper than silver wire. You can find copper wire of many descriptions in the ordinary hardware store. Unlike fine silver wire, which is commonly used for cloisonné work, copper wire will oxidize in the firing process. As the enamel is built up in the cloisonnés, it is necessary to clean the exposed parts of the wire after each firing. The copper color can be maintained after the final firing by employing only acid resistant enamels, thereby allowing a cleansing in a mild nitric acid solution before proceeding with the buffing.

Another variation on the cloisonné technique is to place two wires close together, forming a subtle strengthening of the line. Because the wires are bent freely by hand, the double wire effect is never mechanically parallel, and it adds considerable charm and gracefulness to the design. You must grind the enamel very finely in order to fill the spaces between these parallel lines, which, in many cases, will be extremely narrow.

Mixed heights, widths, sizes; round and square wire; broken, unbroken, and double lines; and the use of both copper and silver wire can be successfully combined in cloisonné work, giving it much more character than traditional pieces, which are sometimes monotonously perfect.

2. *Forming the Plaque*

Although cloisonné enameling can be placed on any surface—whether horizontal, vertical, or curved—it is advisable to form the edges of the

FIGURE 101. Design for cloisonné using ribbon wire 30g. x 18g.

Design for cloisonné using 26g. round wire in short snips.

Design for cloisonné using 16g. square wire.

cloisonné plaque or panel whenever possible. The slight forming at the edges gives richness and body to the work. Perhaps in the case of small pieces when there is to be no stoning and the wires are elevated above the surface, the curved or formed edge can be excluded.

Forming of the panel for cloisonné is done with the chasing hammer or, on thinner metal, with the small horn hammer. Hold the piece at a forty-five degree angle to the steel block (or similar flat metal surface) and by even strokes of the hammer simply cap down, or round, the edges. In the case of a square or oblong plaque the corners should be nicely domed (or balled). File and trim the corners so that the plaque will rest solidly and firmly on a flat surface, touching evenly all around. For a professional job file the edges of the convex surface, and, if needed, follow by the abrasive wheel, then the felt wheel with tripoli, and lastly polish with the cotton buff and jeweler's rouge. This procedure produces a reflective base for silver cloisonné, but in many of the older cloisonné pieces done on copper the surface was left roughened, which supplied a better bond for the enamel.

3. *Cloisonné on Curved and Vertical Surfaces*

On very curved or vertical shapes the roughened surface is of greater consequence. A coarse file or the vibrator tool produces this effect. When doing cloisonné work on the vertical sides of a previously enameled bowl a substantially thicker mixture of gum tragacanth must be employed. Allow the usual tragacanth solution to settle overnight, pour off the thinner part, using only that part which has settled to the

FIGURE 102. Form the edges of the plaque with the chasing hammer on a steel block.

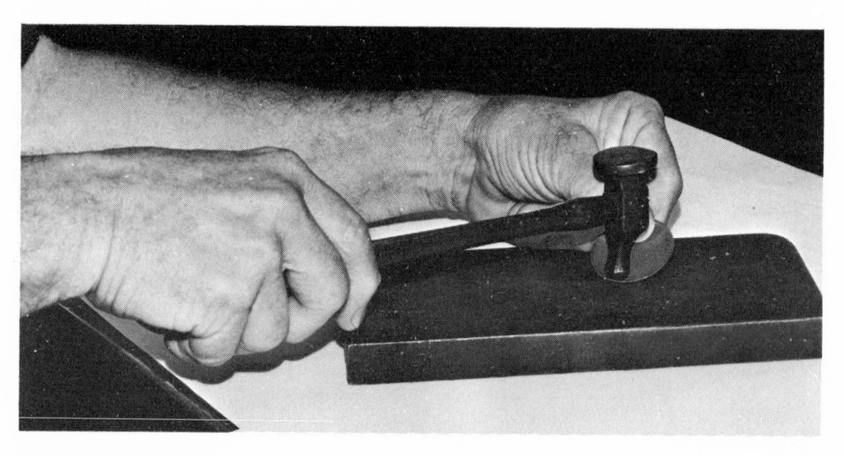

bottom of the jar or bottle. Paint the enameled surface to receive the wires with this thick, heavy mixture and also deposit each wire into it before laying it in place. Permit the tragacanth to become thoroughly dry (preferably by drying overnight) and the problem of the wires slipping out of place should be obviated.

To guard further against wires slipping down on vertical surfaces, avoid soft fusing flux as a base, and fire with utmost caution, never exceeding a temperature of 1450° F. (This is the temperature to be strictly followed in all cloisonné work.)

4. *Mechanically-bent Wires*

Mechanically bent wires may be used in a geometric type of pattern or to supplement free shapes in a cloisonné design. It is easy to wrap the wire (ribbon, round, or square) around a variety of pins, nails, and rods ranging in size from 1/32″ to 1/8″ in diameter. Then slip the spiral of wire off the nail or pin, hold it in the left hand, and snip by inserting the point of tiny manicure scissors. The result will be perfect circular

FIGURE 103. "The Vine," 3″ x 1½″ round covered box by the author; cloisonné and paillons on a curved surface. (*The Cleveland Museum of Art, lent by the artist for the forty-sixth May Show*)

rings. Place them on a steel block, tap gently with the chasing hammer to straighten, and with tweezers close up the gap in the rings.

Other shapes such as squares, rectangles, and triangles can be similarly produced. Make V shapes by bending the wire around a square rod of steel or hard wood and cutting at two corners. The round or square wire is of greater advantage when straight lines are needed. Cut these straight pieces with sharp scissors, or with the metal cutter. Sequences of concentric circles or squares, which produce surface textures, and geometric patterns of infinite variety may be both fascinating and ingenious.

5. *Annealing and Stretching Wires; Tracing*

It is much easier to bend fine silver cloisonné wire after it has been annealed. To do this, place a tightly-bound coil or ring of untwisted wire, held in place by binding wire for even distribution of heat, into the kiln on a clean trivet or fire brick. Keep the door ajar, and with the

FIGURE 104. "Lapis," 2″ x 1½″ x 3″ box by the author; mechanically bent cloisonné wires. (*The Cleveland Museum of Art, lent by the artist for the forty-seventh May Show*)

temperature set for 1450° F., watch the wire until it becomes a light pink color. Have a container of water near the kiln. Plunge the wire immediately into the water; the wire will now be very soft and pliable.

A good first step is to develop an accurate working drawing with fairly strong black ink outlines. Trace this design onto the silver, after first coating the silver with white tempera paint. Use the steel pointer or stylus to obtain a fine black line. Now retrace the line on the silver, using the same pointer and remove the tempera.

The pliable cloisonné wires should be held at one end and formed around each segment of the design as designated by the working drawing. After shaping them, dip the wires into the thick tragacanth solution and then put them in place on the silver plate. Dust a thin coat

FIGURE 105. Mechanically bent wires may be wound around different sized nails and cut with manicure scissors.

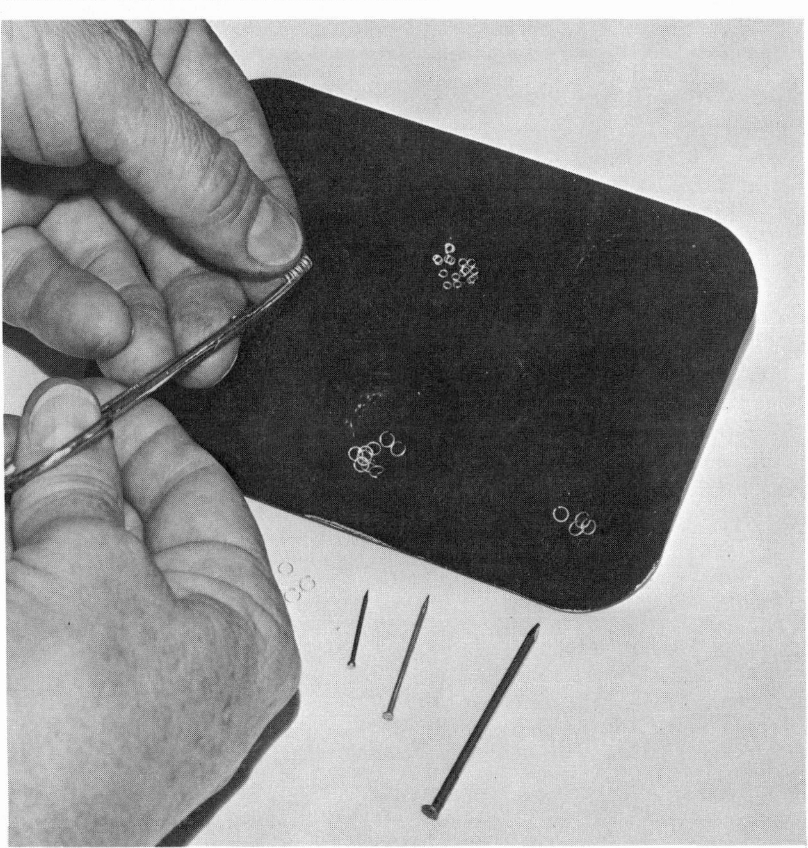

of flux for silver or a transparent champagne over the entire surface, fire, and the wires will be held securely in place.

You can stretch wires by clamping one end in the vise and wrapping the other end around a pair of pliers. Pull with an even, gradual pull in one direction. Anneal again as described above, and stretch again. Obviously, you will be decreasing the diameter or gauge of round wires. If the stretching is done with caution, it is possible to double the length and decrease the gauge by one half. If a change in shape of the wire is desired (round to square, square to flat, etc.) the draw plate must be used.

6. *Colors for Cloisonné*

One of the common mistakes of the student, when first fashioning a cloisonné enamel using silver wires, is to make an unfortunate selection of colors. Because sketches and finished colored designs are often done with a pencil or ink outline, the student is likely to forget that his final colors will be outlined in white rather than black. Light opaque colors (including white opaque, flux for silver, light transparent blues and grays) are often so close in value to that of the silver wire that contrast is lacking, and the design "falls apart." Much of the traditional work (with the exception of the gold Byzantine cloisonné enamels) was developed in strong opaque colors. The use of black, dark blue, deep rich reds and purples made this work strong and meaningful. The aim

FIGURE 106. "Quintus Sanctus" by Linda Woehrman; cloisonné panel. (*The Cleveland Museum of Art, lent by the artist for the forty-eighth May Show*)

to keep in mind is a good pattern with interesting distribution of colors set apart by light outlines. If you leave the wires raised above the surface of the enamel (cloisons purposely not filled nor stoned to a level) having dark colors in contrast to the light color of the wire is of less consequence. In that case the wire casts tiny shadows, which separate each color regardless of its value. However, value contrast is essential in planning a color scheme for cloisonné.

If when the design is completed, the result seems to lack value contrast, there is an interesting method of bringing life and vividness into the finished product. The following procedure is suggested not only as a remedy, but as a kind of "finishing touch" for any well-defined cloisonné. On a flat porcelain tile, grind powdered black over-glaze enamel with oil of sassafras for proper consistency, and add one drop of squeegee oil as a binder. Using a fine crow-quill pen or a bowl-point pen, apply the black overglaze at the edges of each color. Shade the black out into the color areas with a little of the oil, allowing the black accents to run up to the wires but not over them. When fired, the design should have the "snap" or contrast that it lacked. Remember that overglaze must be fired only to the glossy state. A moment of overfiring causes the overglaze to sink, creating unpleasant depressions in the surface.

7. *Free Color Areas*

In spite of the fact that cloisonné was originally a device thought necessary to separate a color from its adjacent color, the modern enamelist need not be bound by such a restriction. In fact, a cloisonné design is often benefitted if the enamelist thinks of the wires not as boundaries for each color, but as integral parts of the design. Accept the wires as lines in the design rather than as outlines, and permit the color areas to pass through or beyond them.

Consider, also, the possibility of using spots or dots of wire for texture or accents. To do this, cut equal lengths of wire (approximately 1/16 inches), place these pieces on a charcoal block, and apply the blowtorch transforming them into tiny balls of metal, or in some cases, strange shaped blobs. These, then, are placed on the enamel concurrently with the wires and attached by a gentle firing. After the enamel is built up around the tiny balls or blobs subject them to the stoning process.

FIGURE 107. Design for cloisonné; it is not necessary to keep colors strictly within boundaries of lines as in traditional cloisonné.

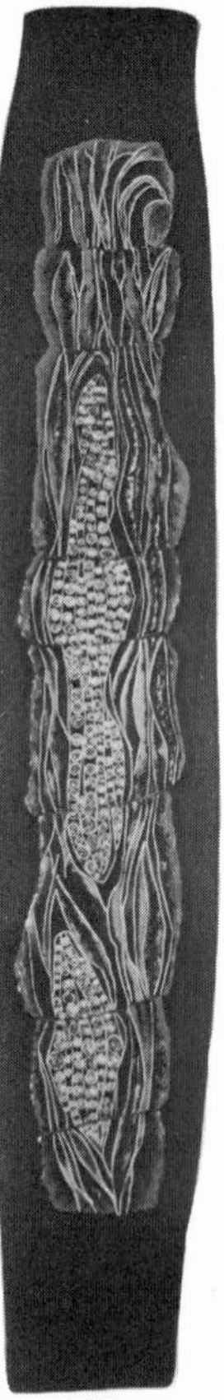

FIGURE 108. "Indian Corn" by the author, cloisonné panel using gold foil within the cloison areas.

8. *Foils with Cloisonné*

Cloisonné on copper as a base metal can be made to look brilliant and sparkling. Being less expensive copper is often more available to the student and, in fact, is commonly used as a base by the professional craftsman. Start the work by forming the plaque as described above, leaving the surface roughened by a coarse rasping file. Counter-enamel the piece and prepare the face by cleaning in nitric acid. Bend and cut the wires according to the outlined drawing, and place them directly onto the copper. Use only a dark shade of transparent enamel for a complete dusting of the surface. (This should be a very thin layer, just enough to attach the wires when fired gently.) Do not use a soft white opaque enamel to attach the wires; during subsequent firings the soft white tends to bubble through around each wire destroying the delicacy of the linear pattern.

Now prepare the silver or gold foils for the various cloisons. Hold the foil between a folded piece of tracing paper, and, starting from the top of the sheet, make horizontal and vertical cuts, producing tiny squares about 1/32 inches in size. Pick up these squares individually with a water-color brush moistened in tragacanth solution and completely cover the surface of each area to be transparent. Keep the foils from touching the wires and fire them in the usual manner. Use soft opaque ivory as an undercoat in other segments of the design, and proceed by filling the cloisons with your most transparent colors, which have been assiduously washed and reground. Fill all parts to one level. Stone, fire, and refill the low places, remembering to refire each time after stoning to remove the scratched surface. Gold or silver wire is equally successful when used in conjunction with a copper base.

9. *Cloisonné on Fine Silver*

There is an added compensation for the enamelist when fine silver is used as a base in place of copper. Although an approximation of the shades and brilliancy of transparent colors may be achieved by the use of silver paillons as described above, the work on fine silver has still another quality—softness or limpidness, a quality unique to enamels.

The accepted gauge for fine silver sheet to be used as an enameling base is Brown and Sharpe 18 gauge or .040 inches. Thinner metal up to 24 gauge is advisable for basse-taille or repoussé work. Either gauge should be adequately counter-enameled to avoid warpage. Sometimes pieces of wire or wire mesh are imbedded in the counter-enamel for added strength. A good rule is to balance the amount and type of counter-enamel with the thickness of enamel used on the face of the plate. Also capping of the edges of the metal as described on page 134 will act as a safeguard against warping or crazing around the edges.

First, immerse the fine silver in a weak solution of sulphuric acid (10 parts water to 1 part acid). This solution (or pickle) should be heated but not to the boiling point. Keep the fine silver in the pickle for two or three minutes. (Little damage is done to the counter-enamel by subjecting it to the acid bath.)

After this process, the silver may be prepared for enameling in one of several ways: (a) Vigorous rubbing in a circular motion with extra fine steel wool produces a satin finish. Before enameling, any oily deposit left from the steel wool should be removed either by saliva or by wafting the surface with a broad blowtorch flame; (b) For an even more brilliant translucency to the colors, buff the silver to a high polish with the felt buff and tripoli, or white diamond compound, following this with a muslin buff and jeweler's red rouge; (c) With the electric vibrator tool held vertically, texturize the surface in a stippled, serrated, notched, or crenulated pattern. This type of surface causes tiny reflections, giving more life to the enamel. The effect is similar to commercially machine-turned pieces that appear deep and sparkling.

After the first dusting of enamel has been fired (preferably a flux for silver, or a flux mixed with transparent gray or transparent champagne), it is essential to stone this to a paper-thin layer and refire before placing the wires. As the cloisonné progresses, often a change of color or shading of tone in subsequent layers adds considerable quality.

FIGURE 109. Surfaces for silver: (1) vibrated surface; (2) steel wool surface; (3) polished surface.

Fill the cloisons slightly above the height of the wires before stoning. Replenish the enamel wherever low or glossy areas appear and continue until the surface is perfectly level.

10. *Gold Cloisonné*

Gold cloisonné or cloisonné using gold (either 24 carat or 18 carat) as a base is not essentially different from copper or silver cloisonné work. In planning his design the craftsman must remember that he is playing his colors against a yellow or orange line rather than a white line as in the case of silver cloisonné. Certain yellow enamels, or colors that have the same value as gold, are, therefore, rather ineffectual. The exquisite gold cloisonné icons done in Byzantium in the late eleventh century often displayed a restricted palette of rich dark reds, blues, and purples as dominant colors.

It is quite possible to proceed with the gold in the same manner as for silver—that is, by firing a thin coat of transparent yellow or flux as a base color and attaching the wires to this by placing them in the kiln. However, with gold we are conscious of a precious metal, one that warrants utmost respect and one that might be used in more intricate, three-dimensional forms, as in the case of jewelry. It would seem appropriate to use some method to hold the wires more securely in place and to preclude the use of a base coat of flux, which in some cases might impair the purity of the color.

Soldering the wires in place, a method used by medieval craftsmen, has two disadvantages: discoloration of the enamels and the risk of melting the solder while firing the enamels.

There still remains another way to attach the gold wires to the gold base plate; it is that of fusing or granulating the wires in place.

11. *Gold Granulation*

Gold granulation has been one of history's major mysteries. For many years after the third and second centuries B.C., when much of the fabulous Etruscan granulation work was produced, craftsmen were intrigued by the process, erroneously labeling it some type of refined soldering. However, under the microscope the tiny balls of gold appear not to be attached by any form of soldering, which would tend to flow under and around them, but to be virtually "perched on the surface."

In 1933 H. A. P. Littledale[1] patented a process of attaching gold

[1] The Scientific and Technical Factors of Production of Gold and Silverwork, London: Worshipful Company of Goldsmiths, 1935.

granules to a gold base, a process entirely devoid of solder. Having discovered that a deposit of copper on the granules or between the granules and the base would, if subjected to the correct heat, allow them to fuse, he developed his theory by simple chemistry. Copper hydroxide and glue were reduced to copper oxide and carbon. The carbon plus oxygen disappeared as carbon dioxide gas resulting in the desired coating or veneer of pure copper.

Based on previous theories and endless research of his own, John Paul Miller[2], a contemporary goldsmith, utilizes the technique of granulation (of which he is a recognized living exponent) to fuse gold wires for cloisonné enameling. Miller is able to fuse the wires to 14 carat gold, thereby giving strength to the piece, by a skillful method that involves reducing part of the gas flame (near the point furthest from the torch). The copper, which must be in juxtaposition with the gold to cause fusibility at a temperature lower than that of the gold, is the result of plating. Miller's process of granulating cloisonné wires is a possibility, but not one for the amateur. After fabricating the parts involved, there is the matter of split-timing when the wires are brought to the same temperature at one time and when the entire surface of the

[2] "Gold Granulation," by Conrad Brown *Craft Horizons,* March 1957, pp. 10–15.

FIGURE 110. "Flounder and Fossil" by John Paul Miller; cloisonné pendant with segments of gold fused to the base plate. (*Museum of Contemporary Crafts, New York*)

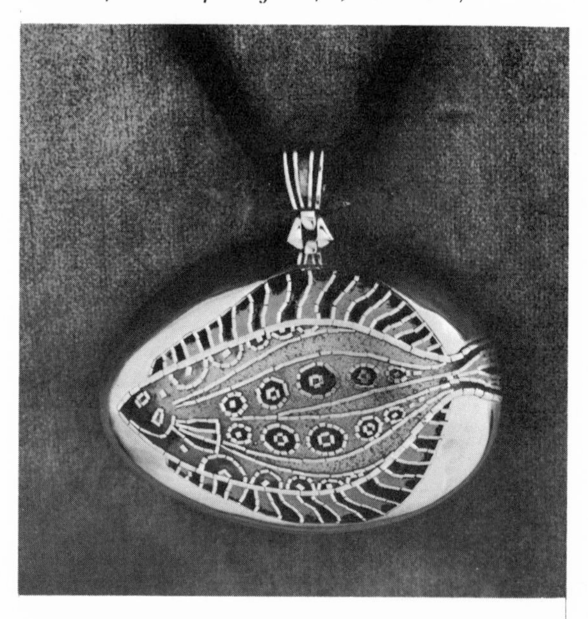

gold seems to become "liquid." Undeniable skill, patience, and dexterity are of paramount importance.

12. *Finishes for Cloisonné*

There is a choice of finishes for cloisonné enamels. The more customary finish is obtained by stoning, first with a carborundum stone of coarse grade, followed by a medium, and then fine grade. Do this under running water. It is extremely important that the stoning be done so that the excess enamel is not "ground down into" the surface, but "ground off"; hence the constant use of water to rinse away small particles of enamel.

The water method is not always followed in the case of larger panels. It is possible to stone large panels with the electric motor, using a fairly coarse carborundum wheel and a slow speed. Of course this grinding could be done with running water, but then both the studio and the craftsman become spattered with a white silt. Instead, carefully scrub the piece, using a toothbrush, ammonia, and Ivory soap. After a thorough scrubbing, the piece is refired with no noticeable ill effects. The use of hydrofluroic acid as a final cleaning agent is sometimes recommended, but I have purposely not suggested this process because of the great danger of the acid to the eyes and skin. It does give an excellent finish and brilliancy to the enamel, easily removing any milky film or deposit of enamel in the tiny pits exposed by stoning. If you ever do use hydrofluoric acid, use it out-of-doors or in a well ventilated studio. Be sure to return the hydrofluoric acid to the container in which it was purchased, and *never* allow it to touch the skin, nor under any circumstances breathe its fumes.

After the surface has been satisfactorily leveled, the piece is dried and refired. This last firing must be watched carefully bringing the enamel only to the molten state. The wires are then polished in the usual manner, first with a felt buff and tripoli and then with the muslin and rouge. All traces of these compounds may be eradicated by dipping a toothbrush into ammonia, rubbing it on a cake of Ivory soap, and scrubbing the surface of the enamel with it.

Another method of finishing cloisonné to produce the professional quality often seen in modern European work is to eliminate the final firing. After stoning with the finest grade carborundum, continue with a Scotch hone. A fine creamy paste will be formed. Use this as you work with the Scotch hone, until all scratches have been eliminated in both the enamel and the wire. The design now has a clear-cut definition, a quality that may be lost with refiring. Start the buffing, but this time use a fine grade of pumice with water or oil deposited on a thick cotton buff. A slow motor is necessary; otherwise the pumice tends to fly off the wheel. The matted surface that was the result of the stoning can eventually be brought to a fairly high gloss, and the sharpness of the edges and clear-cut shapes are therefore retained.

This same effect can be accomplished almost as well with a wide felt buff and bobbing compound, although it is a longer process.

Still another finish for cloisonné can be obtained by leaving opaque black in its matte or stoned state. (For the type of design where color is minimized, the major emphasis and decorative quality of the piece are stated by the wires.) Polish only enough to brighten up the wires, allowing the black to remain completely matted. Other colors could be left matted but are often dulled and less satisfactory than the rich ebonylike quality of opaque black.

These are only a few of the special effects obtainable with cloisonné. It is a flexible technique that allows the individual craftsman ample opportunity for a vast variety of mannerisms, ranging from rough, free interpretations to the most meticulous and exacting executions. Indeed, some craftsmen have found cloisonné challenging enough and absorbing enough to exclude all other techniques in their enameling.

FIGURE 111. "Butterflies and Thistles" by the author, showing value contrasts; gold plated cloisonné round wires were used in this covered box. (*The George Gund Collection*)

CHAPTER 6

POSSIBILITIES OF CHAMPLEVÉ

As long ago as the fifth and sixth centuries B.C. gouged out areas in metal were filled with a vitreous substance and brought to a molten state—a form of champlevé. Later in the twelfth, thirteenth, and fourteenth centuries the champlevé technique was extensively used. All manner of religious panels, crosses, reliquaries, and pyxes displayed this kind of surface ornamentation. Eventually, champlevé enameling was combined with cast heads or figures in the round and semi-round, which were riveted onto the piece after the enamels had been fired. The modern utilization of this technique remains basically the same as it has always been, although stamping and casting of the metal are sometimes done to produce objects such as automobile name plates, class rings, pins, heraldic insignia, and other commercial objects.

1. *Copper Champlevé*

Copper, an inexpensive and available metal and also one possessing color possibilities, is excellent for this technique. To be successful, a design for champlevé must take into consideration the fact that both the negative parts (copper) and the positive areas (enamel) exist separately but that the whole must function as a unit. Without this consideration when developing the design, the result (and this is too commonly seen) becomes nothing more than a meaningless group of disconnected spots. Remember that the metal areas may be separated from each other by a narrow line, if necessary, to bring about a more coordinated pattern.

The usual or traditional method is to plan for the copper areas to be exposed, taking for granted that the copper will eventually become oxidized if not constantly cleaned. To circumvent this, oxidize the copper intentionally with liver of sulphur solution after the enamel is finished. Make a very weak solution by dissolving a small amount of liver of sulphur crystals in warm water, and immerse the entire piece at one time, letting it remain until the proper shade is acquired. Another way to avoid oxidization is to dust a very thin coat of flux over the whole design. Fire this only after the final polishing has been scrupulously carried out.

FIGURE 112. Cross, French, Limoges, twelfth-thirteenth century; champlevé enamel on copper; the champlevé technique was widely used in the twelfth century by the famous Limoges enamelists. (*The Cleveland Museum of Art, J. H. Wade Fund*)

2. *Sgraffito and Painting with Resist*

For the resist a good grade of asphaltum varnish works as well as anything. Some craftsmen prefer to heat this slightly, adding to it a small proportion of melted beeswax. (Actually, beeswax by itself makes an excellent resist; it can be painted onto the copper while hot in much the same way that wax painting for batiks is done.) Apply the asphaltum varnish with a pointed brush at room temperature. Dilute it with turpentine until the negative areas can be painted with satisfactory definition.

Unusual textural quality is sometimes gained by painting the entire area of the copper, and then with a pointer or sharpened brush handle sgraffito, or scratch away, the areas that will receive the enamel. Do this when the resist is partially dry. Such a method produces edges that

FIGURE 113. Design for champlevé box; light gray represents raised metal areas that may be plated.

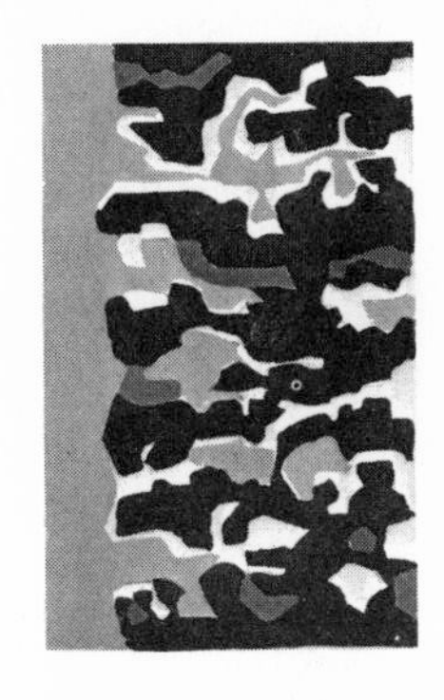

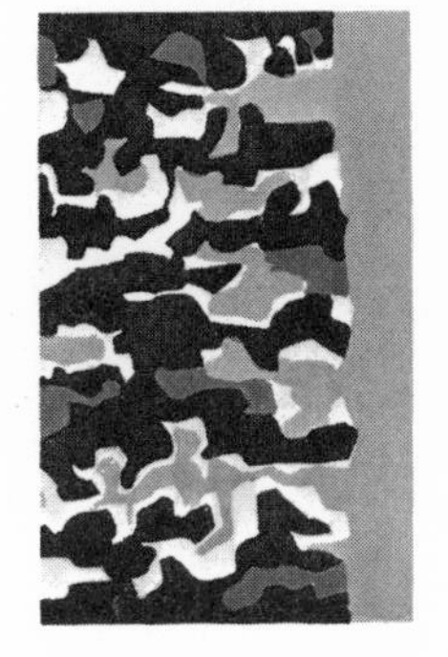

are less stiff and have more character. With the steel pointer you can also sgraffito fine lines in the asphaltum areas. The acid will eat through where the finest line of copper has been exposed. Such an etched line may remain until the final polishing, or it may be filled with very finely ground black enamel.

Retouch all edges, prick and refill any bubbles that may have formed in the asphaltum, dry thoroughly, and immerse in the acid bath.

3. *Depth of Acid Bite*

Etching the copper with acid is much like some forms of cooking, for which it is impossible to give a hard and fast rule. In general: Strong acid bites the copper faster, but has a tendency to "ripple" the edges, that is, "eat under" the edges of the motifs. A strong acid solution will attack copper if there is the tiniest pinhole in the asphaltum and will also disengage the asphaltum if it is not applied thickly enough. A "strong" acid bath is one that creates a vigorous bubbling on the surface of the copper—any solution stronger than one part acid to four parts water might be considered in this category. It is quite possible to control such a solution by observing it constantly and dispersing the surface bubbles with a feather.

FIGURE 114. "Stygian Flames" by the author; gold plated champlevé, with engraved lines on the raised metal "fields."

There are advantages to a "slow bite." To prepare a slow solution, reduce the concentration of acid (nitric) to one part acid and seven parts water. Whereas the strong solution might recess the copper to the proper depth (about $1/16''$ when using 18 gauge, and about $3/32''$ when 16 gauge is used) in less than five minutes, the weak solution can safely be employed for eight to ten hours. You should, of course, have the piece entirely immersed in the bath and, if necessary, weighted down with a piece of heavy wood. With weaker acid solutions the edges will be more clear-cut and pieces of the asphaltum will not be released to float on the surface of the acid.

FIGURE 115. "Vase No. 2" by Jean O'Hara; gold plated champlevé with two "bites" in etching process. (*The Cleveland Museum of Art, gift of the Cleveland Art Association*)

4. *Multiple Acid Bites*

A most attractive and newer interpretation of champlevé is that which has more than one level, that is, double or triple "bites." To do this, proceed as above, wash and dry the piece and cover certain parts of the first bite with asphaltum or wax, leaving the original resist in place, and immerse in the acid bath again. As transparent enamel is applied over the original area, the second and third bites (or depths) give considerable richness to the design. Remove all asphaltum varnish with turpentine or by burning with the blowtorch. *Clean* the copper *thoroughly* before enameling.

5. *Colors for Champlevé*

It should be obvious that colors chosen for champlevé would take into consideration the color and value of the metal, whether it is copper, silver, or gold. Yet many champlevé enamels fail because their designers have ignored this basic rule. Interestingly, more often than not, the older champlevé pieces were composed of opaque enamels. Certainly opaque black, white, primary and secondary colors can be handsomely expressive with copper. One is loathe, of course, to cover a more precious metal such as gold or silver with opaque colors, and yet, by interspersing the opaque colors as minor accents with the transparent enamels in the major areas, it is possible to achieve colorfulness without sacrificing proper value relationships.

Colors need not always blend with the metal. They may be in direct contrast or purposely dissonant. But the satisfactory adjustment of value is basic to the successful planning of champlevé coloring, whether dark on light, or light on dark. An unfortunate choice of color values of the enamel as it relates to the color of the metal will ruin the champlevé piece. On the other hand, when well planned, champlevé can be one of the most expressive enameling techniques.

6. *Plating with Champlevé*

Plating gold or silver on copper, or gold on silver is most rewarding in champlevé work. A certain amount of research and experimentation will be necessary, however, because of the element of heat involved with the plating process. But, generally speaking, plating is much more apt to be successful with champlevé than with cloisonné because the

FIGURE 116. "Alpha and Omega" by the author; cloisonné within champlevé recesses in the manner of the early Byzantine work. (*Courtesy Dr. C. R. Lulenski*)

areas to receive the plating are exposed and consist of one piece of metal.

From the variety of colors of gold, choose that type of gold which combines best with the color scheme of the enameled areas. It is also possible to plate with copper or chromium, although the last might be more appropriate for a geometric or mechanically spun form. In 1925, a multitude of very attractive bowls, cups, and trays were produced at the Kunstegewerbeschule in Vienna, making use of simple geometric bands of opaque blues, black, and white contrasted with chromium plating.

7. *Silver Champlevé*

Working with silver is essentially the same as working with copper. The acid resist (asphaltum varnish), the kind of acid (nitric), and the timing of the acid bite are all similar for silver and copper. The biggest difference is in color. Silver requires playing the colors against a cool, or even whitish metal, and not against a warm orangey-brown color. Certainly, the colorist will be conscious of this difference. Cool colors—greens, blues, and grays—seem to have an affinity for silver, in preference to reds, purples, and yellows. The reason is simple color theory: warm colored metals reflect or "show through" the cool colors and have a tendency to neutralize them, whereas silver, which is gray-blue in color, merely enhances the cool colors.

Wash the transparent enamels, regrind them each time they are used, and adopt only those that were most successful in your tests. Keep in mind that the coarser the grains, the brighter the color. Do not attempt to grind the enamels to a paste as in some cases this causes cloudiness.

FIGURE 117. "Cross with Three Apostles" by William Harper; a fine example of gold plated champlevé enameling.

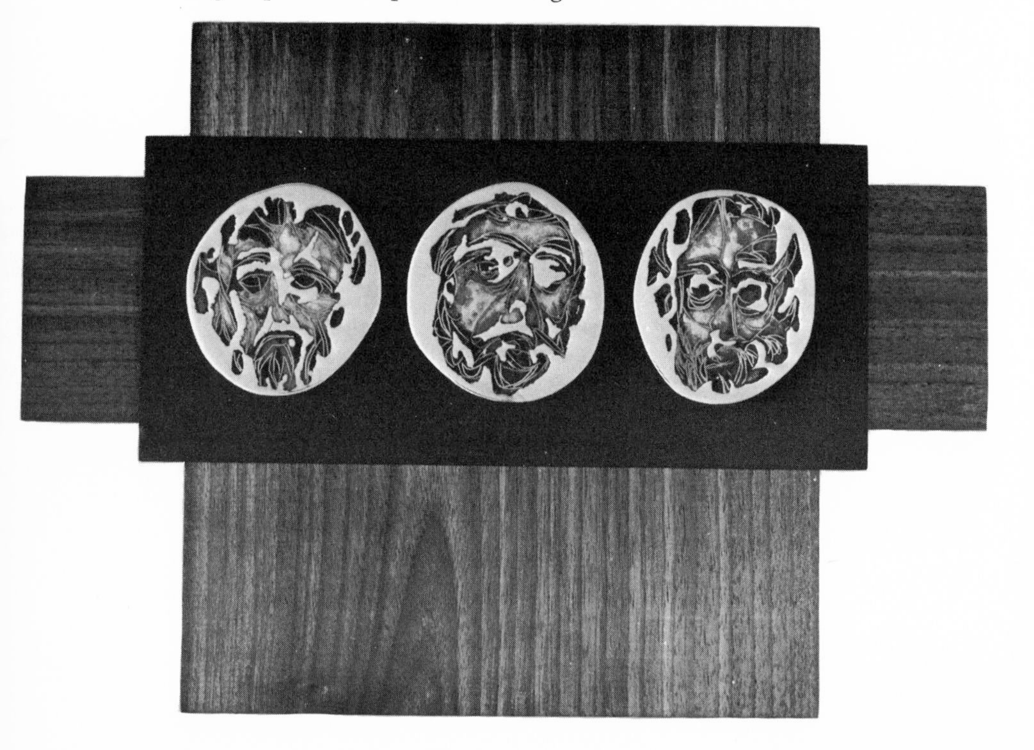

8. *Finishing Champlevé*

Now comes the task of finishing the champlevé piece. It should be stoned with perhaps even greater care than cloisonné. Start with the coarse carborundum, stoning in an even-pressured circular motion. Place a steady board under the slow running tap, allowing its ends to straddle the sink. (This facilitates the chore of grinding or stoning.) Follow the coarse carborundum with the medium and fine stone and eventually the Scotch hone. Keep the water flowing at all times and try to avoid unnecessary deep scratches in the metal. If there are hollows or valleys in the surface of the enamel, you must fill and refire as in cloisonné.

When the stoning is finally completed select one of the three finishes described in Chapter 5 (page 145) for cloisonné: brilliant polish, semi-matte, or matte. Because the metal is exposed in larger areas than in cloisonné, caution must be taken to buff away every semblance of scratches or pits in the metal. Too much pressure with the felt buff will wear away the metal and not the harder enamel, with the unpleasant result that the surface will be wavy. Again, buffing too much in one direction when the stoning has been inadequately performed will result in what may be called "fish-tails," tiny raised streaks at the edge of the enameled areas. To remedy these, simply repeat the stoning in a circular motion, and the buffing with perhaps less pressure on the wheel. This is the time at the final polishing when tiny pits in the metal may appear. Unfortunately little can be done to eradicate these now. It behooves the craftsman, therefore, when doing champlevé to inspect the asphaltum most carefully for air bubbles and refill them before immersing the plate in the acid bath.

A contemporary enamelist should find enchantment in the champlevé technique as he extends it by further exploration and experimentation.

CHAPTER 7

PLIQUE-À-JOUR TECHNIQUES

"Plique-à-jour," which means "similar to a membrane (*plique*) stretched in such a way that the light of day (*à jour*) may pass through," is, in some ways, the most fascinating and intriguing of all the enameling techniques. (One can visualize some fastidious French nobleman in Louis XIV's cortege presenting his Princess with a plique-à-jour necklace that immediately caused wonder and admiration.)

The principle underlying this technique is comparatively simple. It is merely a matter of designing with areas of transparent enamel that have no visible metal backing. Enamel that is at least .04 inches thick is quite sturdy, and when interlaced with wires or small segments of metal having similar thickness, it increases its strength, becoming more durable. What is needed is a method of fusing the grains of enamel while suspended over small areas, and held in place by capillary attraction, or a way to fuse the enamel upon a metal that can then be eliminated. Thirdly, there is the possibility of fusing the enamel and wires together while supported by a metal or other substance to which enamel does not adhere.

None of these methods is too involved, and each presents possibilities for further exploration on the part of the contemporary enamelist.

1. *Sawed Method*

The simplest type of plique-à-jour is developed by the sawed or pierced method. For this you will need to procure a sheet or disk of fine silver which is 18 gauge in thickness. First anneal the silver in the

FIGURE 118. Punch, drill, and saw each hole in this 6″ diameter piece of fine silver for a plique-à-jour enamel.

FIGURE 119. Finished plique-à-jour plate on fine silver by the author.

kiln at 1450° F. until it is pink (about three minutes); then clean it in the usual way by immersing it in the diluted sulphuric acid pickle (one part acid to ten parts water). Keep the silver in the warm pickle for three or four minutes. Remove, dry, and polish the surface with extra fine steel wool, or crocus cloth. Unfortunately, sterling silver is not too satisfactory for plique-à-jour work. Oxidization at the edge of each hole tends to discolor the enamels.

The design for this kind of enameling must be carefully thought out. Each area, though a separate hole in the metal, must give coherence and meaning to the composition. Without coherence, the result becomes little else than a series of meaningless holes. (Here the designer will use his sense of negative and positive area adjustment. Either the enameled "windows" or the remaining silver should dominate.) Transfer the working pattern to the silver plate, and center-punch each tiny area in preparation for drilling. With needle-like drills the center-punching is absolutely necessary. Use an automatic snap-punch, or a sharpened steel point with the chasing hammer.

Ingenuity in creating the design may have included those round holes that can be made by various sized drills. These need no further sawing. After drilling, remove the burr formed on the reverse side. Use the file or finishing stick for this. Sawing is not merely therapeutic relaxation. It must be done with exactitude and perfect control, thereby eliminating tedious hours of filing. Keep in mind when sawing that one hole or shape may be cut as close to the next as 1/32 of an inch or even less without destroying the structural strength of the silver plate. A further development of the plique-à-jour pattern is to insert and solder a network of partitions within the sawed holes. This is usually accomplished with fine silver cloisonné wire, having the same height as the gauge of the silver plate itself.

The sawed holes should not exceed 1/8" for the short dimension and one inch for the long dimension. By grouping many tiny holes close together a peculiar optical illusion is created. When strong light passes through these closely associated holes, particularly when such a group is of one color, the narrow separations or partitions seem to disappear. If the holes are close enough together, the light becomes diffused causing them to appear as one.

Sawing hundreds of holes on a curved plate or bowl is impractical because of the difficulty in holding it firmly, and obviously the saw would hit the opposite side of the bowl. To render a curved form, first saw out the entire tracery while the plate is flat, anneal carefully, and create the three-dimensional shape with a wooden mallet and sand bag. Care must be taken not to injure the delicate network of silver. In

FIGURE 120. Working drawing for plique-à-jour panel showing the use of mechanically drilled round holes.

FIGURE 121. "Plique-à-jour III" with stand, by the author; example of sawed technique with enamel over exposed silver. (*The Cleveland Museum of Art, the Mary Spedding Milliken Memorial Collection*)

working out such a bowl or tray it is of decided advantage to solder on a base ring, enabling you to fire the piece without trivets. Use a hard silver solder that has a melting point above 1450° F. Arrange the design in such a way that the base ring is not noticed from above. Insert the enamel grains in each aperture and proceed with the firing as described later in this chapter (page 166).

2. *Acid Method*

In the acid method the plique-à-jour is made on a copper form, after which the form is disintegrated by acid, leaving only the enameled shell. This technique is a bit more involved than the sawed method, but not too difficult.

Plique-à-jour objects of various sorts and shapes are made in Japan at the present time and exported to expensive gift shops. These pieces are often looked upon as rare objets d'art, notwithstanding the fact that the technique has been known for centuries by both Russian and French craftsmen.

FIGURE 122. Plique-à-jour bowl, Japanese, 1900. (*Cooper Union Museum, New York*)

When finished, the plique-à-jour enameled shell appears to be completely colored; the only metal parts are the fine cloisonné wires. (These wires must be used for structure and as separation for the colored areas.) Here is how to make a small bowl (not over 4″ in diameter and 2″ high) with the acid method. First, realize there are limitations on the design. For example, wires that outline the colored motifs must be close enough together to strengthen the enamel. If possible, they should be "tied together," leaving no large unbroken background areas. Some sort of adjustment must be considered for the base in the form of a heavier wire, or a series of wires, which will enable you to place the piece in the kiln right side up without touching the enamel. A comprehensive sketch in full color is mandatory.

a. FORMING AND BASE COAT. Start by forming the bowl. Hammer the

FIGURE 123. "Perpetual Vine" by the author; cloisonné wires for outside of plique-à-jour bowl.

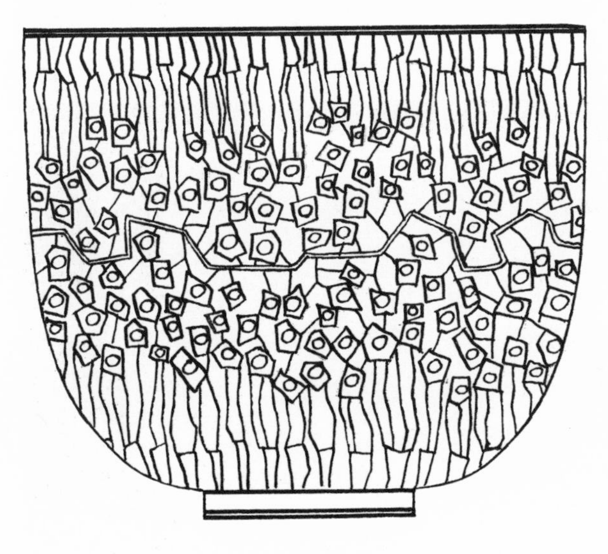

shape from 20 gauge copper (22 gauge copper is sometimes used). As you probably know by now, it is possible to enamel on one side of the copper without counter-enamel so long as you do not apply the enamel too thickly. Clean the copper as usual and apply a thin coat of Scale-off to the concave surface. When this is properly dried, you are ready for the first application of enamel to the convex side. This must be either dusted or wet inlaid and composed of a medium hard or hard, thoroughly washed, clear flux. You may stone, refire, and apply the second coat if necessary.

b. FIRING WIRES. The silver cloisonné wires are bent and set in place by heavy tragacanth on the fired flux surface. They are attached by a gentle firing. With too much firing they will sink into the flux, which must act as a protection against the strong acid needed for disintegration of the copper. By using gold cloisonné wires the whole procedure would be altered inasmuch as gold is impervious to nitric acid. In that

FIGURE 124. "Perpetual Vine" by the author; in this plique-à-jour bowl the inner cup of copper was completely dissolved by nitric acid. (*The Harold T. Clark Collection*)

case you could immerse the whole bowl in the acid as long as you retain a tiny rim of copper or capping of gold at the top by asphaltum resist. Such a rim is needed to strengthen the otherwise fragile glass bowl.

c. FILLING PARTITIONS. Now proceed by filling the wire partitions (cloisons) with enamel. Only the most transparent (by testing), washed, and reground colors should be utilized for this purpose. Purple, dark greens, and dark browns are not clear enough for plique-à-jour. Apply the colors evenly with the spatula, tucking the grains into every detail of the configuration. Fire at about 1450° F., stone, refire, fill the low areas, stone, and fire again until the surface is even and completed. Be sure that every hole is flooded over with enamel. Replenish the Scale-off at each firing to avoid bits of fire glaze damaging the enamel. Polish the wires and finish the edge in preparation for the next step. More will be said about firing for plique-à-jour later (page 166) in this chapter.

d. ELIMINATING THE COPPER BOWL. The next step is to eliminate the

FIGURE 125. Bent silver cloisonné wires are placed on the *outside* of the heavily fluxed copper surface for a plique-à-jour bowl.

copper bowl itself, leaving only the enamel and wire shell. First paint a thin band of asphaltum varnish about 1/16″ wide around the inside, top edge of the bowl. This rim, when polished, gives a better finish but, more important, it is needed to provide strength. To disintegrate the copper, place the enameled bowl in a larger Pyrex container and pour a strong solution of nitric acid (approximately equal parts of acid and water) into the smaller bowl, filling it only to within 1/16″ of the top rim. The acid must not touch the outside of the enamel as it would immediately affect the exposed silver wires. Stir constantly with a feather to aid the reaction of the acid on the copper. Have a second Pyrex bowl at hand to receive the acid as it is poured off, as well as a large bowl of clean water for rinsing.

It is important to work in a well-ventilated studio—or preferably out-of-doors—as the brown fumes caused by the reaction of the strong acid will be overpowering and dangerous to breathe. With wood tongs, gently lift the enameled bowl from its Pyrex outerbowl, pour off the acid, rinse, and check several times until every trace of copper is removed. If the action is too severe and too rapid to control, reduce the strength of the acid solution and "eat" the copper away more gradually.

FIGURE 126. The plique-à-jour bowl is swabbed with concentrated nitric acid until the copper bowl itself is entirely disintegrated, leaving only the shell of enamel supported by the silver wires.

3. *Non-adhering Materials*

Of the three materials to which enamel when fired does not adhere, the most commonly used is high-fired sheet mica. (The other two are aluminum bronze and brass.) Only the smooth-surfaced type of mica serves the purpose and, in most cases, a tiny, quite transparent film is deposited onto the enamel. However, if the mica is not old and flaky and is kept intact, you can be assured of complete separation when the piece is cooled. Note that sheet mica is used only for firing *flat* pieces (which will be discussed later, page 166).

To make a plique-à-jour bowl using a non-adhering metal (from which the bowl will separate when cooled) you must use either aluminum bronze or brass. I do not recommend the first metal because of two reasons. (1) It is extremely hard and not malleable enough for forming in the usual way. (2) There are a variety of proportions in the alloy of aluminum and bronze. For instance, I have experimented with a proportion of 68 per cent aluminum and 32 per cent bronze with little success. Bronze, which is an alloy of copper and tin, is also smelted in numerous ways containing different quantities of copper. The presence of too high a proportion of copper in aluminum bronze rules it out as a perfect non-adherent.

Even though brass is also an alloy of copper and zinc, it constitutes the best source of non-adherent metals for enameling. (Some form of brass may be seen in Oriental or Medieval enamels, but only rarely and, then, only with opaque enamels.) To make a small plique-à-jour bowl, first form the shape from 18 gauge brass. Do not anneal it, if possible, as this deposits an oxide on the surface that is apt to permit the adhesion of enamel. After hammering the desired shape, polish the brass on the convex side to a high gloss. The matter of polishing must be stressed, because it aids non-adhesion. Place the fine silver, rectangular ribbon, and cloisonné wires on the outside of the bowl. When filling these with transparent colored enamels take care to avoid the surface of the wires. Apply a slightly heavier wire at the base and rim. Use fine silver for these and solder together with hard silver solder. Clean and polish the joinings of the wire rings. Cover the soldered joints with powdered rouge or ochre when firing. Apply the carefully washed transparent enamels rather heavily, completely filling each cloison. Now fire the bowl upside down at approximately 1500° F. or until all enamel has become thoroughly matured. Firing in the reversed position is also possible. If all has gone well, the inner brass bowl should separate from the enameled shell when cool.

The inside surface of the plique-à-jour bowl will be very smooth but discolored by the oxidization of the brass. Immerse the bowl in acid (1 nitric acid to 4 parts water) having the same room temperature as the temperature of the enamel, and let it remain until the enamel is clear. If desired, this entire process with brass may be reversed—that is, you could polish the concave side of the bowl and apply wires and enamel to it rather than the convex side.

In all forms of plique-à-jour enameling the wisest rule to follow is that of gradual cooling and avoidance of drafts. Place the piece, still on the hot trivet, on top of the kiln—or, even better, let it remain in the kiln—turn off the switch—and remove when cooled.

4. *Firing*

Firing plique-à-jour enamels is a more subtle and delicate process than that of firing enamels onto a metal base. Here the concern is to suspend the moistened grains into small openings (as described on page 163) with capillary attraction helping to hold the grains in place. (Sometimes a temporary base in the form of mica acts as an aid when the enamel is fired.) When firing a flat piece of plique-à-jour, the pattern, which is composed of a pierced sheet of silver or a group of wires soldered together, should be laid on or clamped to a sheet of high fired, smooth mica. Upon completion, the plique-à-jour should lift off the mica sheet, leaving clear transparent colors in the openings.

A problem arises however, when the shape is curved. As sheets of mica do not bend, some other method is necessary to hold the moist enamel in place. The solution to the problem is another form of mica. A natural silicate derivative, it can be divided into thin, tough laminae (or scales). (These pieces, known as isinglass, were used extensively in old-fashioned lanterns.) They are transparent to translucent and are available in many shades and densities.

From a sheet of mica peel off the thinnest slivers imaginable. Break or cut pieces into about 1/4″ squares. Dip these in a tragacanth solution and stick them to the back of the shape, covering over each sawed hole. When dry, they will create a perfect foil for the enamel, prohibiting it from running through the apertures. With this method, fewer firings are required.

It is still possible, however, to fill the holes with enamel without the use of mica pieces on the back—by using capillary attraction. Plique-à-jour enamels must be fired in one position and then reversed. As the enamel is liquid or runny when molten, the firing requires some skill and precision. Enamel will tend to draw toward the edges of each hole on the first or second firing. These must be filled time after time, using the principle of capillary attraction until the hole is eventually "flooded over." To minimize this kind of pulling away in the firing, simply fire at a lower temperature. As with all enameling on fine silver, there is a tendency of the enamel to "crawl," especially at first if too high temperature is used, but because fine silver does not oxidize in the kiln, such defects are remedied in subsequent firings.

With the bowl shape, both the ring at the bottom and the rim at the top enable you to fire by reversing the positions alternately. This alternate reversal will negate a heavy drip of enamel at the rim and also excess enamel at the base. At all times fire only until the enamel is fused and remove from the kiln immediately, but you can expect to have a great number of firings. (It is not uncommon to fire plique-à-jour enamels as many as twenty-five times, reversing the position each

FIGURE 127. Small segments of mica are placed on the reverse side of the plique-à-jour holes to retain the enamel while firing.

time until every hole is finally filled.) The surface can be stoned, refired, cleaned, and buffed in the usual way. Keep in mind always that silver is softer than enamel, and when the two are in juxtaposition as in this case, you must buff gently; otherwise the silver areas will appear sunken or worn down and the enameled areas raised.

5. *Ways to Use Plique-à-jour*

Historical examples of plique-à-jour enamel include elaborate necklaces, chains, and dangle earrings made in Italy in the sixteenth century. Also well known are finely wrought and exquisitely enameled demi-tasse spoons, made by Russian craftsmen, that appear to resemble miniature stained-glass windows when held to the light.

Although plique-à-jour is less practical than other techniques, it should not be ruled out because of that. "Practicability" is not the most flattering term to be used regarding a piece of art. Perhaps the plique-à-

FIGURE 128. Gold bowl with plique-à-jour enamel by Margret Craver, enameling by Earl Pandon, 1955. (*Museum of Contemporary Crafts, New York*)

jour, if well designed, is justified in being an "objet d'art," a delightful bibelot to "have around." There are several ways of displaying plique-à-jour enamels. A tray, dish, or bowl is often best shown in an upright position, perhaps on a small silver stand designed so that only points touch the back of the plate, allowing a maximum amount of light to pass through it. A deep bowl is sometimes displayed on a small box with a frosted glass top, which contains a miniature electric bulb that lights the bowl from below. Deep shadow-boxlike frames around the plique-à-jour can also be made to contain small lights. Hung in a sunny window or placed on a side table where light can penetrate them, plique-à-jours usually play the role of conversation pieces.

There is yet another possible use for the "backless enamel": contemporary religious (or secular) light fixtures might be designed to accept plique-à-jour motifs in the form of insets or panels. For larger areas of plique-à-jour it is possible to use copper screening, place it upon a sheet of smooth mica, fill the tiny holes between the mesh, fire, and remove the mica. As explained earlier (page 165), polished 17 gauge brass, being a non-adherent to enamel, may also be used for such a purpose.

This particular phase of enameling, where the craftsman is deliberately concerned with light passing "through," rather than "into" or "on to" the surface, adds one more facet to a fascinating medium.

FIGURE 129. Larger holes may be sawed from a piece of metal for plique-à-jour enamel if it is fired to copper screening, which automatically reduces the size of the holes to 1/16″ squares.

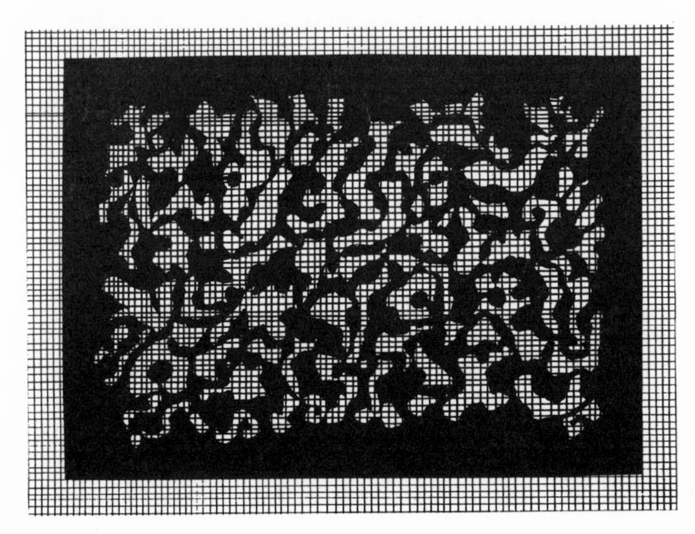

CHAPTER 8

PANELS

The last decade has brought a sudden and unprecedented demand for enameled wall panel plaques and small pictures. Utilitarian objects such as enameled trays, boxes, and plates may be considered too expensive, or too hazardous to be within the reach of small children, whereas something hung on the wall is more practical and at the same time becomes a status possession. For whatever reasons, at the present time vertical, horizontal, square, or round enameled panels to be used as wall decorations, either singly or in groups, constitute a large proportion of the enamelist's endeavors. By a "wall decoration" or a "decorative panel" I do not mean to imply an illustration, a landscape, or a realistic picture. (And never should the traditional carved picture frame be used for an enameled plaque. This is an appalling bastardization too frequently seen in many hobby shows.)

1. *Working Drawings, Sketches*

My constant harping upon the matter of making preliminary color sketches for enameling is perhaps annoying, but I would be more than remiss not to make some mention of it in connection with planning a panel, where you must make decisions about the shape and size, the position, whether it is to be horizontal or vertical, and the way it is to be mounted. (It is not practical to make these decisions in metal or wood; therefore it behooves you to "think on paper.") Decisions like those and also ones regarding color, subject matter, style, and technique are seldom reached before the creative artist has used up many sheets of his sketch pad.

The manner in which you are going to present the idea—painted, cloisonné, champlevé—will influence your thinking. The aim could be for any of a number of effects (hard outlined edges, soft diffused dusting, kaleidoscopic intensity of color, and so on), but such decisions must be formulated *before* you cut the copper. Enameling is not a medium that can be erased or washed off like certain art media although, in one sense (the fact that it can be refired or reclaimed) enameling does have considerable flexibility. The enamelist can recover the entire surface with an opaque enamel and start afresh, or he may crack off the enamel with a metal hammer and steel stake saving only the copper for a second attempt. However, if he had made the decisions in the first place, such expedients would be unnecessary.

An artist has a right to work in his own manner, and there are those who disagree with the preliminary sketch theory—who prefer to work directly into the medium—but it is my contention that freedom of expression need not be hampered, and is actually aided, by preconceived resolutions. As with any medium, and more so with enameling, which is subject to endless variations upon becoming fused, the artist must never allow himself to be enslaved by his original sketch, but use it only as his guide in a general way. On the other hand, he should

FIGURE 130. "Rattlesnake" by Linda Woehrman; a 12″ x 18″ panel with gold paillons and outlines of black crackle enamel.

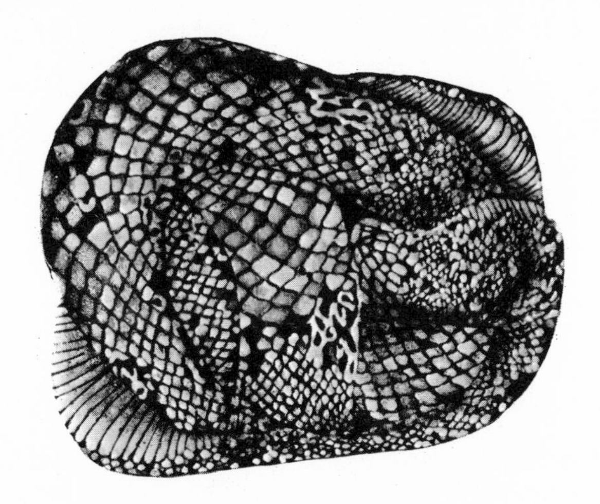

strictly follow the more mechanical aspects of his plan (thickness of wood, working drawing for segments, gauge of metal, placement of brazing pins, and so forth).

2. *Fitted Sections; Sawing; Firing*

As the painter may choose between stretched canvas, gessoed board, water-color paper, etc., so the enamelist in creating a wall plaque or panel may choose from a variety of materials and methods. First, there is the sawed and fitted sectional panel, a method that allows you to develop a panel of any size so long as each part does not exceed the dimension of the kiln. Also to be considered is the possibility of a large mural composed entirely of individually fired enamel sections. These sections could be interspersed with areas of wood veneer of the same thickness as the enameled copper. In that case first mount the veneer

FIGURE 131. Sketches for the vertical panel "Camelot" by the author were done quickly and freely.

sections in their respective positions and then the enameled segments.

Here are a few hints. In the original planning of the design, cuts for the subdivisions of the whole panel should follow along edges of colors, stem lines, or anatomical structure lines. These should be so chosen that the breaks in the pattern are unobtrusive and become an integrated part of the drawing. (Making these cuts at random with little consideration for the pattern results in an amateurish look.) Do not attempt to subdivide the panel into complicated or sharp pointed shapes. Expansion of the copper when the individual segments are fired disallows such shapes to fit together. The most helpful suggestion is this: Saw all component parts of the panel with *one saw cut* and

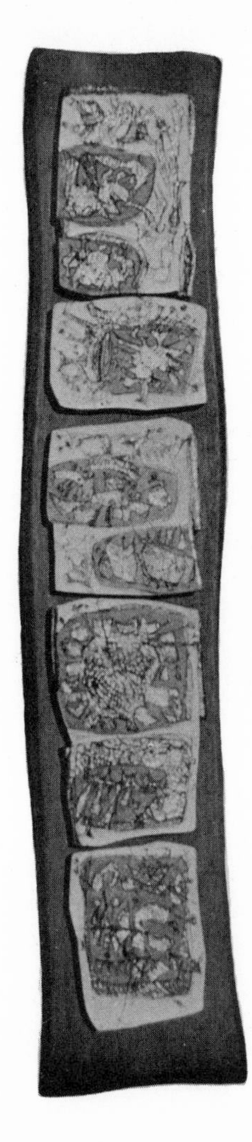

FIGURE 132. "Camelot" by the author is a vertical panel 30″ x 8″ composed of separated units. (*Courtesy Syracuse University, N.Y.*)

from the *same piece of metal.* If you trace the sections on separate pieces of metal and cut them individually, fitting them after they are fired becomes an exasperating chore.

Pieces of the jigsaw puzzle must be kept absolutely flat. To do this, first fire a good coat of counter-enamel on the back of each piece, then brush Scale-off directly onto the fired counter-enamel, and fire this into the surface. Upon the surface of a flat piece of Marinite asbestos or similar type of firing plate, paint a thin coat of Scale-off, firing this in also. Now when the segments are enameled on their front sides, they may be laid directly on the firing plate, placed in the kiln at any temperature, withdrawn, and separated with no adhesion whatsoever. Pieces fired this way will remain flat throughout the remainder of the work, particularly if they were straightened between the counter-enamel firing and the first face firing.

When the panel is put together and mounted, a small amount of grouting may be added between the segments. Commercial grout for tiles, plaster of Paris, Savogran, or any of the wood filler products are satisfactory for this purpose. The grouting must be skillfully applied, stained, or colored in such a way that it becomes quite invisible. Guard against heaping the enamel near the edges of the sections, and stone them assiduously to create the illusion of one level, unbroken panel having few, if any, highlights.

FIGURE 133. "Forest and Trees" by the author; detail of enamel panel with wood inlay showing wet transparents over hard white technique.

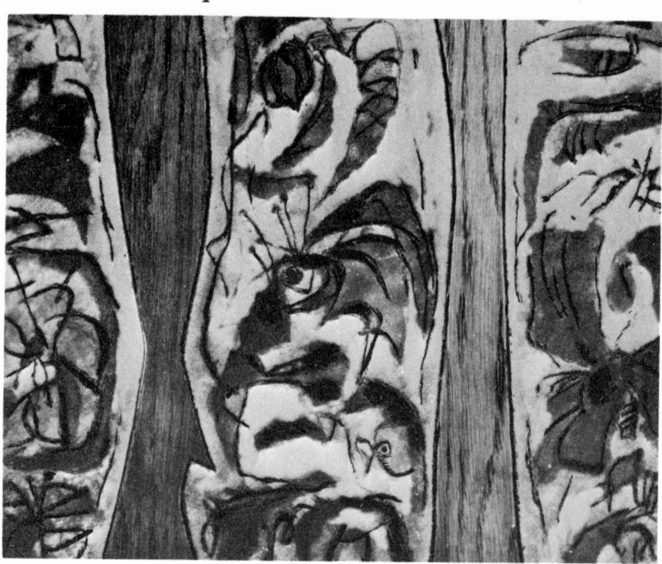

3. *Overlapped Sections; Three-dimensional Panels*

The next method of developing a large panel (by "large" is meant anything bigger than the interior capacity of your kiln) is by overlapping sections. In this method, the original design is not necessarily subdivided in the same way, i.e., each segment could be a unit within itself. Such a panel might be made up of a series of small rectangles, squares, or interrelated free shapes. The edges of each of these should be filed and finished nicely by buffing. (Perhaps a very slight curving at the edge of each part gives a better effect than does keeping the parts absolutely flat.) After they are mounted, the polished edges of these units should be blackened by oxidizing with a liver of sulphur solution applied by a small brush.

The intention with this type of panel is to present a more three-dimensional effect. Cut pieces of 1/8″ to 1/4″ plywood or masonite board the same shape as each section respectively, but 1/4″ smaller in size. By arranging these various thicknesses of plywood backings, you are able to achieve different levels and also to overlap the component parts of the panel.

It is essential that the adhesive employed (Pliobond or an epoxy) makes good "bond" with the enameled sections. As with the fitted

FIGURE 134. The working drawing to "Midsummer Night," a 30″ x 8″ composite panel by the author, shows how the five sections are cut on existing lines of the design.

FIGURE 135. "Midsummer Night" was planned to conceal the five sections of which it is composed. (*The Erdelac Collection*)

sections discussed previously, you would need to fire the Scale-off into the counter-enamel, leaving a fairly rough surface. Attaching the plywood back pieces to the wood panel is no problem: there are numerous contact cements that are reliable when sticking porous to porous, or for larger murals you can nail or screw pieces to the panel. The edges of the pieces should be well sanded and, as seen from a side view, they should be stained in a color matching the wood of the panel. However, in sticking the enameled plaque to the plywood raisers you will be confronted with adhering porous (plywood) to non-porous (enamel plaque). Here, make sure by tests that a permanent bond has been made. If the adhesives used are in any way questionable, confer with the local supplier of adhesives. (This is a field involving modern chemistry. The yellow pages in any locality should list some companies that will gladly solve the problem. By stating your case regarding size, materials, and conditions these companies are sure to have the answer. In these days of modern research, it is safe to assume that anything can be stuck to anything with a lifetime guarantee.)

4. *Forming Single Panels; Large Panels; Matte Finishes*

The copper that is to be used as a single panel (one occupying most of the interior floor of the kiln) should have greater forming or "capping" at the edges to add strength and to keep the plaque from warping. (Forming is explained in the chapter on cloisonné, page 133.) When the larger piece of copper receives its first counter-enamel firing, there will be considerable warping. Using asbestos gloves, hand bend the piece while it is still warm. Favor the convex side by bending it up slightly more than it appears to need. In subsequent firings (and particularly high firings) this forced forming will be compensated. The final shape of the plaque should be level and capped at the edges

FIGURE 136. "Metamorphosis" by the author is composed of three parts for ease in firing.

rather than sunken in the middle. Stoning the single plaques, making use of a large carborundum stone of coarse to medium grade will be most rewarding. Ripply surfaces on these plaques, which cause unpleasant reflections and highlights, destroy detail and negate the true quality of enamel.

Sometimes an enamelist desires to make a bulged or protruded surface by repoussé from the back of the plate. To avoid disturbing highlights it is necessary to eliminate the shine or glare. Because hydrofluoric acid is dangerous and impractical for such large pieces, a

FIGURE 137. "Dandelion Fuzz II," a 12″ x 15″ panel by the author showing capping at the edge of enamel plate.

better way is to dust with an even coat of matte glaze and fire at a low temperature. Fortunately, matte glaze does not destroy the color pattern when it is cautiously applied in several thin applications.

If the larger panel is to be framed rather than mounted, less attention to forming the edge is required, but when it is to be bezeled, the edge demands perfect craftsmanship. If the panel is made on previously coated steel, which is available commercially, capping at the edge is less essential because of the rigidity of steel. Extremely large panels have been accomplished with enameling (page 129) in which case there is no forming or capping as these are used architecturally and are incorporated into the structure of the building.

5. *Forming and Composing with Welded Section*

The parts of a decorative panel can be shaped in a method other than sawing or cutting; they may be created by the welder's torch. In other

FIGURE 138. "Ocean Depths" by the author; panel mounted on mahogany showing dusting, gold crackle and white overglaze.

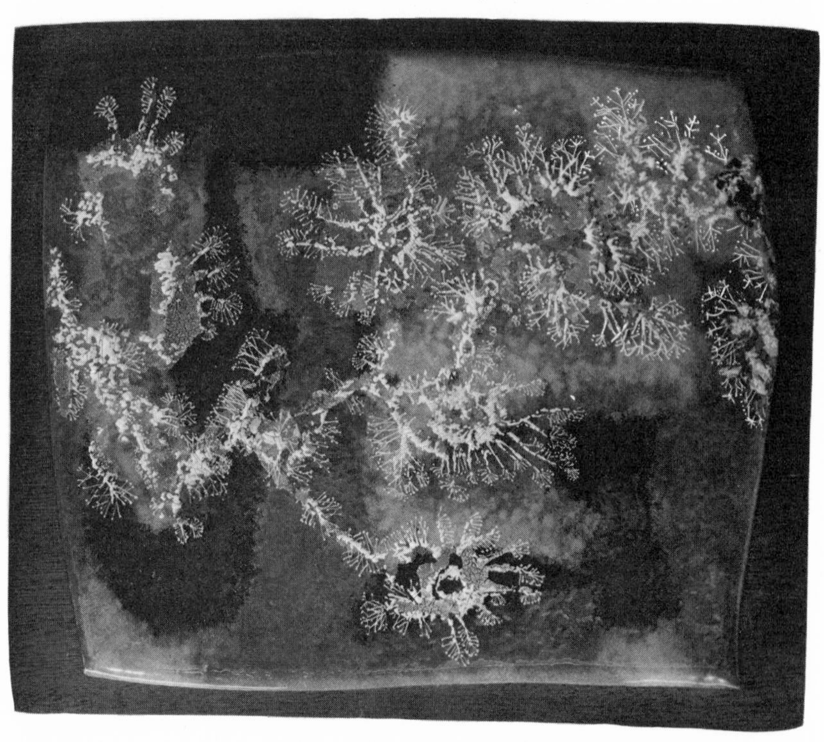

words, with a flame hot enough to melt copper, the shape can be burned out.

Welding implies the fusing of two pieces of metal of the same kind, one to another, while molten, (i.e., when brought to the same temperature). Welding differs from brazing, which is the fastening together of two pieces of metal by a third metal (an alloy in the form of a rod). The alloy is melted at a temperature much lower than the melting temperature of either metal. Brazing creates joints that are stronger than the metals themselves.

Common to both welding and brazing is the source of heat—the oxygen-acetylene torch. Sometimes a "Prest-o-lite" acetylene tank is sold by welding distributors, but more commonly known are the separate oxygen and acetylene tanks. Each tank is equipped with pressure gauges; the gauges and the hose pipes that connect to the torch are clearly marked in red for the acetylene and black for oxygen.

Various nozzles are available. The best nozzle for this method acts as a cutting torch. Oxygen goes through the center of the flame, which approximates a heat of 4070° F. Larger tips such as #6 give forth the most heat, but #3 suffices for our needs.

By regulating the proportion of oxygen to that of acetylene, copper can be puddled and flowed like any liquid. For the panels to be

FIGURE 139. "Last of the Red Hot Poppies" by William Harper, showing burn-out, sgraffito, dusting, and spatula techniques.

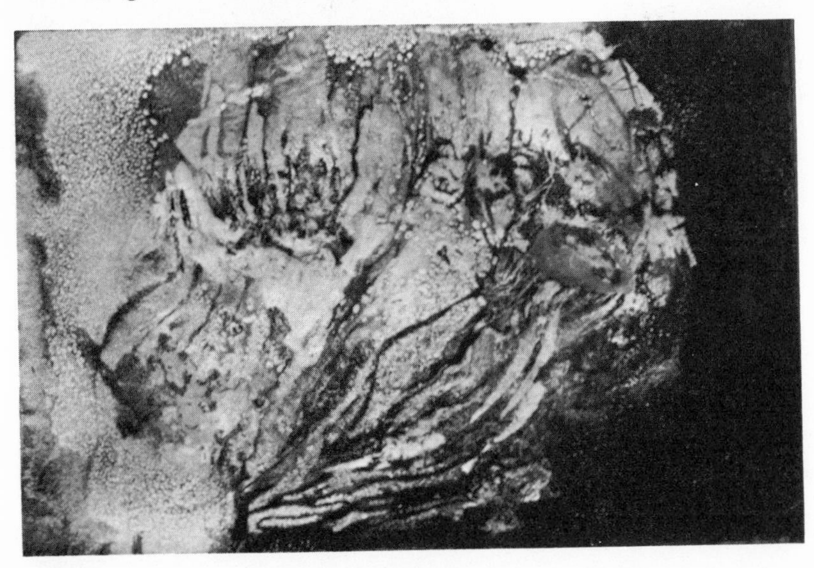

enameled, use the cutting torch (acetylene). After cutting the shapes roughly with the tin snips, or heavy shears, pass the flame, which has been reduced to a white point at the end of the nozzle, along the edge of the panel until a beading is formed. This produces an interesting rippled edge. With the same cutting torch, you can burn more fanciful edges, or create holes in the copper sheet by allowing the flame to remain at one point only moments longer. Protect your eyes with goggles at all times while working with the oxy-acetylene torch. The flint gun is used for lighting the torch after balancing the proportion of oxygen and acetylene until the flame is in perfect control (i.e., producing its maximum heat). Approximately three parts oxygen to one part acetylene can be used, but this does not constitute a hard and fast recipe. You need to experiment to arrive at a satisfactory solution.

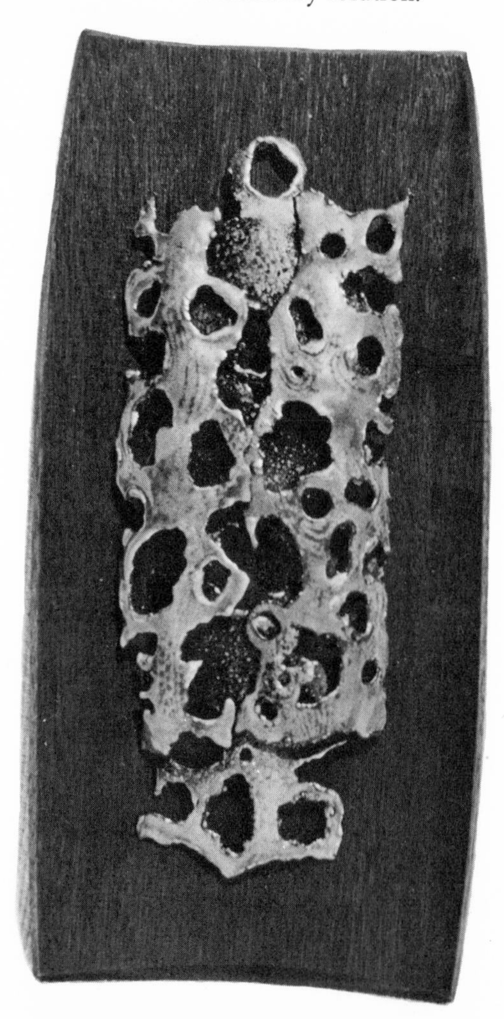

FIGURE 140. "Sycamore Bark" by the author; a composite welded enamel that takes its title from the source of inspiration.

FIGURE 141. "Grenadine" by the author shows the result of the welder's torch applied directly to the copper. (*The H. O. Mierke Collection*)

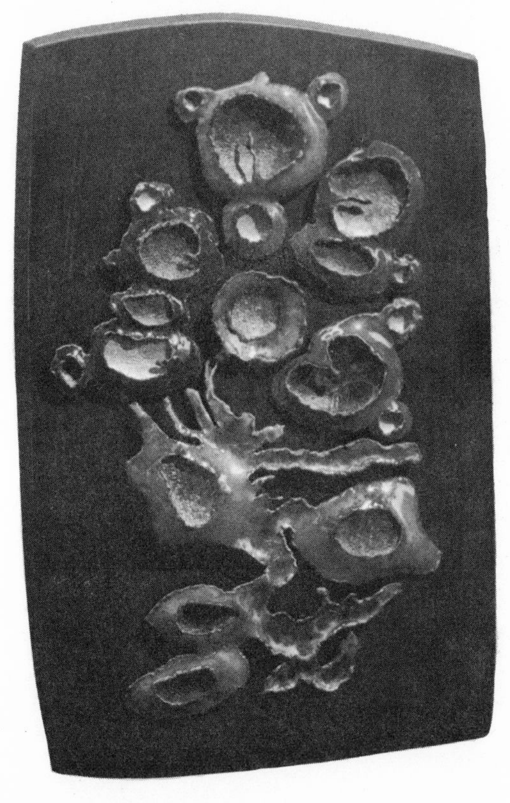

FIGURE 142. "Potted Plant" by the author; welded sections attached with brazing pins; enamel applied to concave and convex sections.

6. *Brazing Pins*

Another way to attach the panels to a mounting board is to use what might be called "brazing pins." Attach pieces of bronze rods, varying from 2 to 3 inches long and about 14 gauge in diameter, to the back by brazing. The melting point of bronze (1868° F.) is rather close to the melting point of copper (1981.4° F.) so by making use of a flux containing borax the melting point of the bronze rod is somewhat reduced. A long piece of the brazing rod may be held with the bare hands. Dip the end in the powdered flux (borax) and touch to the copper at the desired point of attachment. After sufficient heat from the torch has been applied to both the rod and the copper, the rod will melt first. The result is a joint with greater strength and tenacity than one produced by any form of soldering. Interestingly enough, enameling with normal temperatures (1500° F.) may be done after the brazing pins are attached. Reversely—brazing the pins after the enamel is fired—would, of course, be impractical, and the enamel would be destroyed. The enameled sections or panels with attached pins are now ready for mounting.

FIGURE 143. Detail of "Crowd in Subway" by the author; a panel 14″ x 32″ composed of welded sections. (*Courtesy Henry W. Blazy*)

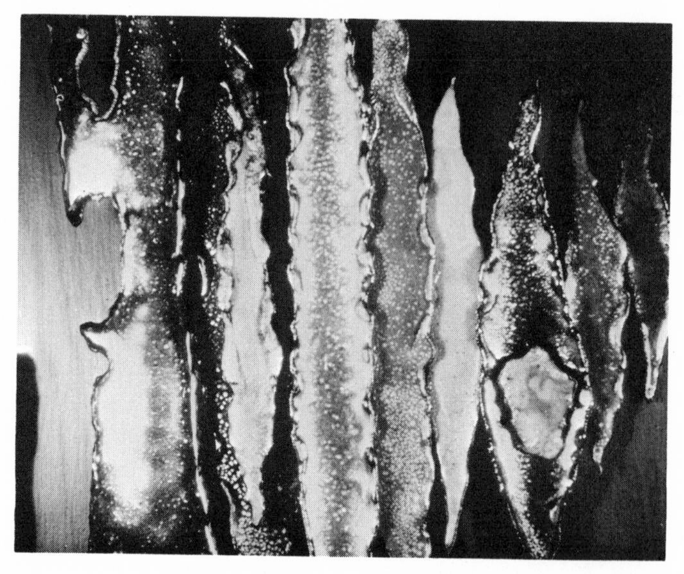

7. *Finishing Welded Edges*

In the case where the panel is not framed or capped over by a moulding of any sort, an exposed edge can be anticipated. As described above, it is possible to create a rather fascinating effect by playing the welding torch backward and forward along the edge of the copper, causing it to bubble up or become rippled. There are various other ways of treating this edge: Because the copper is left in an oxidized (or very black) state by the heat of the acetylene torch, it must be cleaned before enameling. Use a strong acid bath (one nitric acid to two parts water) for this cleaning, allowing the action of the

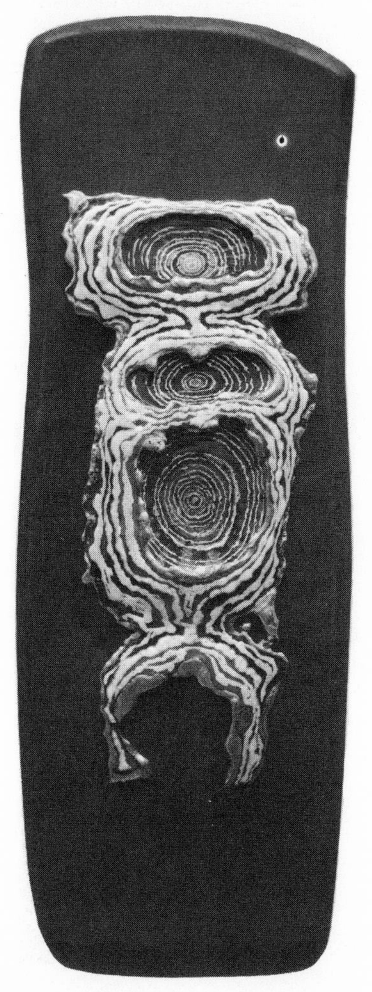

FIGURE 144. "Striped Insect" by the author; welded form with reverse insets and gold plated edges. (*Courtesy H. O. Mierke, Jr.*)

acid to turn the copper a light pink. Buff the edge vigorously with a large muslin buff and plenty of tripoli. (The large muslin buff polishes the uneven surfaces caused by the welding and reaches into the deep crevices.) As the heat of the torch has also softened (annealed) the copper, if desired you may form it or repoussé it at this time.

One idea is to enamel by the wet-inlay method only up to the polished, rippled edge; then gold plate it after the enameling has been completed. Another plan might be to dust a thin coat of flux over this edge, thereby retaining a bright copper color. And thirdly, you could expose the edge, burning the oxidization to a jet black for a free, or more casual look. Holes may be burned into the panel itself, the shapes raised around the holes, other pieces bent reversely, inserted from the back, and held in place by epoxy.

8. *Mounting Panels; Woods; Finishes*

At the time of the preliminary sketching you should give serious consideration to the color, thickness, shape, and size of wood to be

FIGURE 145. "Festival in Byzantium" by the author; a panel 12″ x 18″ making use of gold and silver paillons and translucent colors.

used for mounting. Possibly the enamelist's most difficult and most trying task is not the coloring and the firing of a panel but the problem of how to mount it—how to present the piece in its final state. Too much supplementary mount or frame is as disastrous as too little. Usually a serious artist experiments with cardboard, trying different shapes, sizes, and colors until he finds a solution that "feels" right to him, at least. I don't think any artist should be definitive about a mount without experimenting. Usually the problem has more than one solution.

The science of picture framing is too involved to discuss in its entirety here, but in brief, it should be said that enamels are not necessarily pictures and should not be framed as such. They are better thought of as decorative plaques or panels and are more effective presented in a deep shadow-box, or thrust forward on a block that is spaced away from the moulding, or mounted separately on a piece of solid wood. The wood shape, whether rectangular or free, should be so related to the enamel panel that it enhances rather than detracts from it. Any of the woods with a very prominent grain (particularly after staining) are usually a bad choice. The richness of enamel colors and

FIGURE 146. Detail of panel "Noonday Sun" by the author; shows use of line sgraffitoed through to the original copper plate.

textures call for a wood approximating a flat simple tone when finished—one that is not too "busy," with no capping or framing at the edge. Any type of plywood requires a great deal of filing, sanding, filling, and finishing, and at best lacks the quality or refinement of solid wood. For coloring, experiment with a variety of wood stains, or mixtures of several; allow some to penetrate and others wipe off immediately with a soft cotton rag. For light wood finishes, first paint a flat white undercoat, which should dry overnight. Continue with colored stains, diluting them with turpentine if necessary. Fine steel wool helps to blend the stain and bring out certain accents.

Stains have one bad quality—they affect other surfaces that come in contact with them over a period of time. A panel that has had stain applied to the back of it will discolor the wall upon which it is hung if not protected in some way. To do this either spray several times with a transparent plastic spray (such as Krylon) or cover the back with a piece of good quality felt.

The artist must choose among sharp, curved, beveled, rounded, or free edges for the wood backing. The "doing" of any piece of craftwork is never as difficult as "what to do." Because knack is conceded to be the craftsman's stock in trade (or he should never have chosen such a profession), it is assumed that the mechanical saws, sanding machines, carving tools, and files will respond to his demands.

The peculiar concern for "feeling," which is so evident in the work of mature potters, weavers, jewelers or enamelists, can never be entirely supplied to the student either by teachers or books. It is absurd to think it can. Unless the student has some natural endowment, some faculty or aptness for craftwork as such, there is little, if any, hope for progress.

CHAPTER 9

APPLIED DESIGN

Sophistication about art is not as abundant as some lecturers would have us believe. Mediocrity and conformity are very much with us. It often shocks an artist when browsing in expensive shops to come upon the kind of applied design that is for sale. These are the objects which are presumably purchased for the better homes. There is no assurance that an object of higher price is in better taste than a cheaper one. Wall plaques with gilded Oriental dancers, ceramic ash trays with dripping glazes in gaudy colors, bizarre lamps, and ornate furniture exist in high-priced shops as well as in dime stores. Only an innate sense of taste and appropriateness allows one to be selective in a way that surpasses monetary evaluation.

Improvement of taste by association and through practice is one of the major reasons for attending advanced art classes. Undoubtedly, the beginning art student today approaches his first class having had much more exposure to the current trends, as a result of accelerated curricula in secondary schools, than his counterpart of several decades ago. Strangely, however, although he professes sophistication, the problem of creating a simple shape, deciding upon a color for it, or applying a motif to it, becomes as burdensome to him as it was to his unsophisticated counterpart. Had he come from an environment where good taste in magazine advertising, automobile design, architecture, television programming, or city planning were at the Utopian pinnacle that they claim to be, further education in art would be quite unnecessary. But such perfection is yet to be achieved in our society.

1. *Function and Decoration*

"Applied design" is a misnomer inasmuch as it does not mean "applying a design" to an object. More correctly, the connotation is that a sense of taste, proportion, and rightness were part of the designer's thinking when he created that object. If the object has function and also is in good taste, then texture, color, and shape would undoubtedly have been an integral part of the designer's planning. Therefore another designer can never successfully superimpose his idea of decoration to that piece or that object. The words "decoration" and "ornamentation" are fighting words to most true artists, even though the layman delights in using them—perhaps unknowingly as a torment. He speaks of "decorating" a vase or preferring a screen door with more "ornament" on it. A vase might have a textured pattern that would make it more decorative than one without texture, but that is not the same as painting a bunch of pink roses "onto" it as an afterthought.

Decoration in its better sense (meaning that which is added to, or superimposed upon, a basic shape or form to enhance its interest or content) may be analyzed as follows: (one) structural type of decoration or that which is built upon the structural lines of the form, such as radii, axes, or concentric lines; (two) nonstructural decoration, meaning motifs not built upon these concentric lines or axes; (three) free distribution of motifs existing within structural boundaries; and (four) motifs, lines, or textures that are placed arbitrarily upon a form. In each case the decoration must, without exception, be that which speaks to and about the form upon which it is placed. It can never be foreign to it and must not defeat the shape, but enhance it.

Ornateness is still another term, and one difficult to pinpoint. When does an enameled box become ornate (overdecorated) and when does it escape this fault? Should the artist answer this question or should the art critic, and why should the opinion of the person who desires to own the piece always be deprecated? The integrity of the craftsman himself—not the critic's opinion, not the layman's taste—must be the basis of judgment. Inasmuch as the craftsman was the creator, with him alone rests full responsibility. He should develop a sense of discrimination and selection regarding his work before it leaves the studio. Too often the selection is left up to a jury. Because some craftsmen are wont to "fall in love" with every piece they make and lack the courage to discard what they honestly know to be inferior, they submit pieces other than their best efforts and expose themselves to harsh criticisms. By avoiding selectivity the craftsman has done only half of the job.

Purely organic decoration resulting from chemical and technological manipulation of glazes and fusing (such as textures, break-throughs, burn-outs, and overfirings) are undoubtedly those having the most meaning and those about which the enamelist should show more concern.

2. *Contemporary versus Dated Styles*

To pinpoint "contemporary" style would be meaningless because whatever mannerism or stylistic handling is in vogue at the present writing would be outmoded before this book goes to press. We might better talk about that style, or that manner, which is "significant of our times." It is a concern of those who work in the crafts to grasp something that is the reflection of our present culture. Surely, pat, stiff, sentimental shapes and colors do not relate either to our advanced scientific era nor to our unsettled sociological society. Our lives in this period are at times twisted, torn, fragmented, and disturbed. How, therefore, can art in any of its forms fail to portray something of this fomentation, if it is to be an interpretation of that which exists?

In the fields of painting and sculpture, sweet, tender colors and soft, smooth modeling have long since become outmoded—having been replaced by vital and often times savage interpretations. This change in style is true of the crafts as well; pottery, weaving, jewelry and enamels (to a lesser extent) have become more amorphous in shape and more expressive as the craftsman explores new potentialities of his medium. However the narrow line between freedom of expression and

FIGURE 147. A seventeenth-century English wall decoration by Grinling Gibbons. (*The Cleveland Museum of Art, Grace Rainey Rogers Fund*)

crudity is one the craftsman dares not disregard. Tortured, maltreated metal or dirty, inferior enameled surfaces do not necessarily constitute freedom of expression, no more than the blind copying of trite, modern design motifs and mannerisms makes the work contemporary. A craftsman would be more honest with himself to base his design inspiration on some fine historical style than to adapt only the superficial appearance of contemporary art instead of its essence. To adapt trends as they appear on the scene, with no personal issue or no conviction other than to be in style, is faddism. And yet the craftsman must have a sense of awareness of life about him, of social and economic structure, and of the way people feel and think. This awareness and reaction to the world he lives in can still be expressed in his work without jeopardizing his personal message and convictions.

It is imperative that any performing craftsman have some knowledge

FIGURE 148. In contrast to the preceding photograph, this handsome sculpture by Richard Lippold reflects contemporary taste; it is made of gold, silver, copper, and nickel. (*Virginia Museum of Fine Arts*)

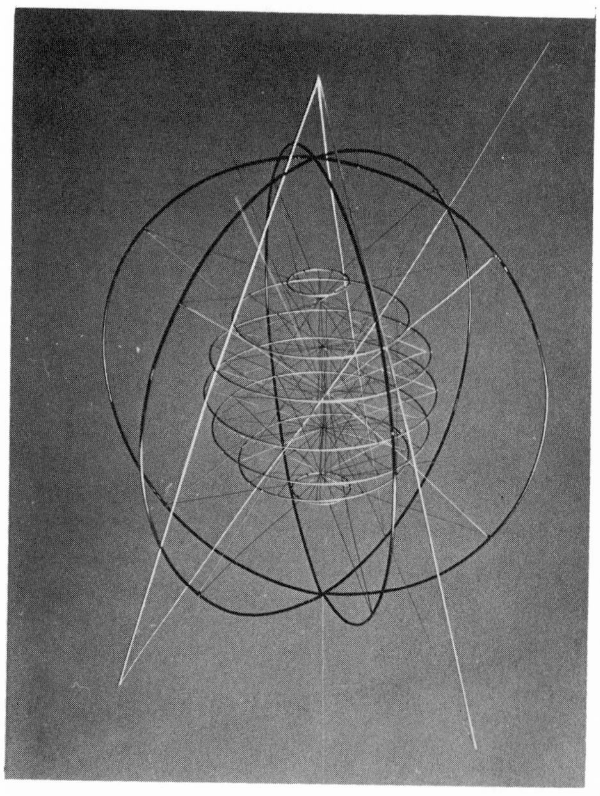

of past stylistic mannerisms. (How else would he be able to avoid a "noncontemporary" look to his work?) Surely scallops, curlicues, little dots, and forget-me-not flowers are not as interpretive of our present culture as more fractured and amorphous forms. The prevailing attitude and taste today (it is hoped) precludes sentimental tendrils, rose buds, and sweet-faced cherubs as applied design motifs. A brief review might be as follows:

In 1930 a kind of geometric style was prevalent. Furniture, lamps, costume accessories, and wallpaper to be in high style often portrayed the skyscraper motif. The work of Piet Mondriaan, a Dutch artist (1872-1944), influenced the fashions of the day. It is interesting to note that in 1966 dress design revived some of this same geometric style.

In 1935 the Austrian Applied Arts School (Kunstgewerbeschule) had a strong influence on style. This style was not entirely geometric, but included also a peculiar hard-edged juxtaposition of stylistic leaves, classic profiles, and abstract flower forms.

In 1940 shapes became more wavy and a sort of amoeba-like motif was seen everywhere. This shape, not unlike the shape of a cashew nut (or perhaps an ear or an oyster) became so prevalent that it soon

FIGURE 149. "Standing Peacock" by John Steck; an interpretation of a peacock in a contemporary manner. (*America House, New York*)

FIGURE 150. The geometric style of 1930 sometimes reflected the skyscraper motif.

FIGURE 151. In 1935 a strong influence known as the Austrian Kunstgewerbeschule style became prevalent.

FIGURE 152. The amoeba shapes were a stylistic influence of the 1940's.

FIGURE 153. Boomerang and lobster claw shapes became overly used in the 1950's.

developed into a cliché. The wilted watch in "The Persistence of Memory" by Salvador Dali (1904-) undoubtedly had a part to play in the popularization of that shape.

In the 1950's the amoeba shape gave way to a kind of sharp pointed motif. Designs for every purpose reflected what might be termed the "lobster-claw" shape. Sharp points and elongated boomerangs appeared everywhere. Enameled ash trays, men's neckties, printed textiles, and dinnerware inevitably became involved with this shape in its many variations. Alexander Calder's mobiles, which had shapes similar to these but beautifully refined, added to their popularity.

Then in the 1960's the enamelist and those working in the decorative design field became impressed with the school of painting known as "abstract expressionism." This form of painting sometimes made such radical use of paint as running, dripping, or splashing. So, because of the influence of such masters from this period as Pollack, Motherwell, and Soulages, craft designers began to adopt a similar style. The reflection of purely organic shapes, colors, and textures, the result of serious exploration with various media and also an expression of the

FIGURE 154. Abstract expressionism in painting influenced designer craftsmen in 1960.

craftsman's inner convictions, seems to constitute our present trend. There are as many styles as there are artists, and the enamelist in creating a unique piece is justified in thinking that his way of doing it constitutes that which makes it contemporary.

3. *Conventional Style*

Because a design is developed in a conventional or symmetric manner does not mean that it is less imaginative than one that is not. Much of the art of primitive civilizations and ancient cultures was dominantly conventional, and much of it is revered by connoisseurs and recognized by museums as superb examples of design. On the other hand, conventional design, which often makes use of such elements as spirals, stiff multiple outlines, and frets, can become monotonous and trite when utilized without imagination or feeling.

It is a misconception to believe that the student of design must avoid all use of the deliberate balance of equal components in order that his work will be labeled contemporary. Rules set up by insensitive peda-

FIGURE 155. The conventional style is often symmetric.

gogues forbidding the matching of two similar parts in a design, by flower-arranging instructors who taboo "two of a kind," or by the interior decorator who will never permit equal spacing of furniture in a room have little, if any, validity. If we were to base all of our theories about design on nature alone, we might argue that perfect symmetry is rare. The tracery of branches in a tree, two blooms on an amaryllis plant, the radiating petals of any composite type of flower, or two sides of the human figure and face are never quite the same.

Of course, if "conventional" is used to mean following a trend or a particular style, this, then, becomes abhorrent to the artist—it becomes regimentation, a most distasteful condition for him who esteems liberation. The creative mind searches for nonconformity, for the unconventional. On the other hand, if we use "conventional" to imply that type of design that is evenly balanced or duplicated in exact repetition, then it seems illogical to rule out this style from our vocabulary.

4. *Abstract and Representational Styles*

During the eleventh, twelfth, and thirteenth centuries, in particular, enameled objects of both religious and secular nature were used as a direct means of communication. The stories of the Old and New

FIGURE 156. "Fox Tails" by the author; tray with three sections hammered from one piece of copper, an example of modern use of spiral design.

Testaments were portrayed on ecclesiastical objects to be consigned to the house of God. Such religious teaching by means of picturization were done in an illustrative manner that would be indisputably comprehensible to the illiterate peasants. Nothing vaguely unrealistic would have served such a purpose. During the Early Christian and Byzantine eras symbolism and inconography were employed, but such symbols or idols appeared in a representational style only.

The modern enamelist also reveals an expression of life in his time, but in an entirely different manner. He narrates turmoil, excitement, cynicism, or *joie de vivre,* as the case may be, by means of a nonrepresentational (abstract) interpretation. Perhaps communication by means of abstract design has even more vital impact (message) than the literal one. Who could say that a huge black area at the top of a Motherwell canvas revealing only a light streak at the bottom of the picture does not successfully get its message of oppression across even to the most untutored observer, or that the tortured distortion of the bull in Picasso's "Guernica" does not reveal the devastation and futility of war? Abstract communication concerns itself with the artist's interpretation of content. (Abstract essence is not the aim of the person who takes snapshots, but may well be the result—and often is—of the study of photography as a fine art.)

FIGURE 157. An example of nonrepresentational style, which may include both abstract and nonobjective.

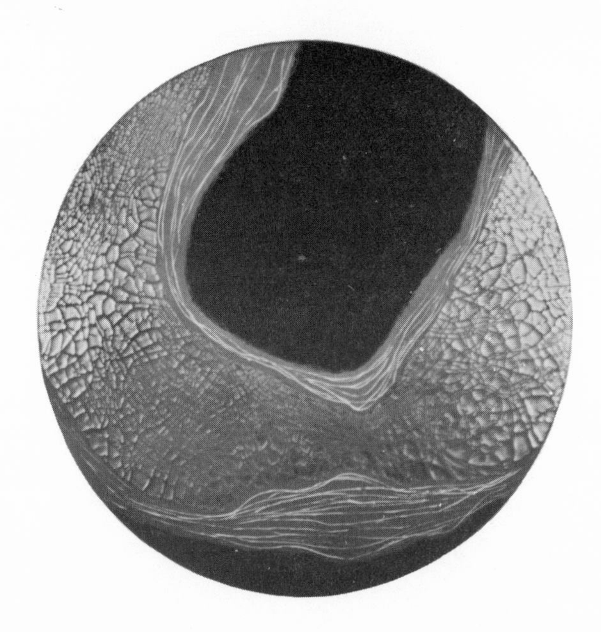

The enamelist must be consistent throughout a composition. Literal facts rendered in a representational way dissociate themselves from the abstract interpretation. But, also, once the factual style is started, the whole design should be completed with convincing realism. To do so requires research and knowledge of every detail.

The illustrator or realistic artist is wont to say that what he does is more difficult than the abstractionist. This argument holds little water as both skill and imagination play the same role in either type of work. It might be conceded that some artists are more inclined to think abstractly than others—some more concerned with detail and realism, or some more introspective and poetical. A piece of creative craft work is never a question of easy to do or hard to do; it is more a question of being inspired enough to want to do it well. The layman inevitably asks, "How long did it take you to do it?" and the artist is seldom able to give an answer. The number of hours involved in expressing what he has to say is of least concern to the craftsman. He may eventually have to be more realistic when selling his creations, but the time element was certainly not foremost in his thoughts.

The medium of enamel is well suited to abstract interpretation. Enamel, which is flowing, moving, fusing, running, changing when molten, is a natural medium for creating illusions. Like any form of glass, enamel reveals depths, imagery, and fantasy. As lights play upon it and within it, the inner composition, which is best expressed abstractly, changes, becoming oscillatory. Flooding the colors while moist, making use of overfiring, break-through, sgraffito, and endless textural effects are but a few of the techniques so well suited to abstract rather than the realistic manner of working.

5. *Op Art and the Geometric Style*

People in the art world are repeatedly being asked the question, "Is this or that school of thought here to stay?" In the sixties the question became, "Is op art here to stay?" Logically, how can one predict what will happen in the future? Every school of thought in the realm of creative art was "here to stay." By that, I mean, it was an integral and important step in the history and progress of art development. Dadaism, impressionism, post-impressionism, expressionism, abstract expressionism, pop art, op art, and top art existed as ways of thinking and seeing. Usually they had been instigated by talented artists who were willing to digress from that which had already been done.

As a form of interpretation the geometric attitude is not new. Piet

Mondriaan, Josef Albers, and Victor Vasarely have long been established as promulgators of this kind of perceptualism. Optical art, which is both a psychological and technological conviction, does prove that visualization on the part of the artist changes throughout history. The present day op artist is more seriously concerned with intellectual and retinal attraction than artists of certain earlier periods. He is none the

FIGURE 158. "Walking Shadows" by Julian Stanczak; an example of op art.

less attempting to prove the fact that "seeing to knowing" can be as deceptive as a magician's act—that the intellect does not always give a correct report of visual experiences.

Op art, an embodiment of the geometric style (with compositions made up of complicated frets, networks of bands, radial arrangements, squares, rectangles, circles, triangles, diagonals, concentricities, moires, and new relations between shapes and colors, both symmetric and assymetric) abnegates many former beliefs. Because these rigid op art compositions are made by ruler, compass, spray gun, and masking tape, the belief that personalized brush strokes, attractiveness of individual techniques, or mannerisms of color diffusion constitute observer appeal has been entirely abandoned. The op artist claims pure objectivity, that the impersonal approach is more valid than the personal, that he has no desire for the observer to feel any sense of security and

FIGURE 159. "Counterchange Flowers" by the author; a 12″ plate that reflects some of the illusionary principles used by op artists.

that he, the creator, the innovator, would prefer to remain anonymous. It might be argued that whether using individualistic brush strokes or flat mechanistic areas of color, it is not possible (and not intended really) for the artist to conceal personality. Only the op artist's technique becomes impersonal; his conception, arrangement, and coloration remain substantially personal.

The artist working in a geometric manner, and particularly the op artist, is searching for new experiences in color, perceptual responses, and semantic imagery. The following titles clearly demonstrate the validity of such a statement: *Degree of Vividness* by Richard Anuszkiewicz; *Four Colors Around White Vision* by Max Bill; *Dynamic Optical Deformation of a Cube in a Sphere* by Enzo Mari; *Ulterior Images* by Julian Stanczak; *Blue White Ford* by Edwin Mieczkowski; and *Study of Differentiation and Identity in Visual Perception: Three Variations* by Lily Greenham.

It is absurd to think that the enamel craftsman, whose work often consists of the fabrication of three-dimensional objects (both utilitarian and purely aesthetic), should feel that he must adopt each movement

FIGURE 160. "Oriental Casket" by the author; a 2″ x 1½″ x 3″ box showing mechanically bent wires combined with paillons and burnished silver details, an example of applied design.

in painting as it comes along. The theories being propounded by op artists (easel painters) are not necessarily those that can be translated or transformed to function as part of the design for an enameled bowl or box. If the enamelist is using his medium on flat surfaces and thinking of it as a form of painting, his use of the principles of op art might be more reasonably justified. Fortunately in our age of individualism no one school of thought in art ever completely dominates the scene, nor should it. Our present American society is too heterogeneous, having adopted many cultures in a short period of time, to produce any one style as preponderant, let us say, as was art nouveau at the turn of the century.

The enamel craftsman need not, therefore, become involved in op, pop, or top art. He may respect these theories about painting, but unless he discovers that they present a means by which he is able to express in enameling something peculiar to that medium, he should show no further interest in them. His contemporaneousness will not be assured by completely forfeiting the intrinsic "look" of enamels, which, by its very nature, is concerned with accidental fusability and soft textural effects, and in its place suddenly adapt the hard-edged geometric conformations of op art.

If anything, to be contemporary is to search, to explore, and to pursue an honest conviction. The conviction that enameling is still in its infancy after many centuries and that new effects and new uses for the medium are yet to be uncovered is indeed a healthy one, and constitutes, in a larger sense, what is meant by "applied design."

AFTERWORD

Everyone who creates, it would seem, is justified in revealing more about himself. This need not be for egotistical reasons, but, for those who are interested, it would make his work take on added meaning—it would reveal his purpose for working the way he does, and his preoccupation with certain phases of his existence.

Mainly because I hope other artist-craftsmen will be more revelatory about their background (and also because I wish to justify a deep-rooted preoccupation that I have with nature), I shall endeavor to set down a few autobiographical paragraphs.

At a very early age, in the small town of North Scituate, Massachusetts, much that may have appeared insignificant to others affected me immensely. There was a small pond, only about ten feet in diameter, with a rock beside it—which became part of my world. Into this small pond flowed a tiny rivulet, which in turn was fed by a constantly flowing underground spring. I don't remember that the water flowed out from the pond, but it must have, although my boy's mind concerned itself least of all with logic. My deep pleasure (and today it remains as a clearly defined visual image) was the indescribable transparency of the rivulet that fed the pond. I spent days on end staring into this clear-as-glass water, imagining all manner of fantasy and reveling in the always changing visions that could be seen below the surface. I thought I saw in that stream all of life before me, what I would be at 25, 30, 50, and even 80 years of age. I sat in utter silence, except for the quiet rippling of the stream and an occasional towhee's leaf ruffling, contemplating life some eight decades hence.

Around this, and other ponds, including Luther's mill-pond, were wild flowers of every description. These became almost like friends to me. Their personalities were more familiar to me than many of the townsfolk. With the wild flowers I had no arguments, no pretense. *Lobelia cardinalis,* or cardinal flower, was so red and velvety that I could almost see below the surface of its three downward thrust petals. If the arrowhead or the cow-lily happened to produce a rarely seen blossom, I thought it was for me. Many, many rare and seldom seen blooms such as the painted trillium, fringed gentian, arethusa, and

FIGURE 161. Tiny and inconspicuous weeds often develop seed pods of amazing style and grace.

trailing arbutus were revealing themselves to me, I was sure, because no one else tramped so many miles just to pay them homage.

Otherwise, my childhood days were spent predominantly on the farm. My grandfather was my exemplar. I thought of him as one of the finest craftsmen who ever lived. Everything about him, his planning, his tools, and the way he used them revealed indigenous and superb craftsmanship. There were sickles, scythes, flails, and pitchforks, all of which were so beautifully balanced and so respectfully cared for that any one of them could have qualified as a worthy entrant to the most discriminatory modern craft exhibition. His heads of cauliflower, of uniform size and of immaculate white, were so skillfully trimmed and so accurately fitted (four to a bushel box) that to me they represented a work of art—a fine piece of craftsmanship. It was he, I remember, who, recognizing my great love for flowers, would stop old Dan, sweating and hitched to the mowing machine in the very center of the

FIGURE 162. "Romantic Garden" by the author; one often remembers flowers that were associated with certain childhood experiences.

west field, merely to avoid mowing down a magnificent clump of black-eyed-Susans. He also must have recognized their startling designs. Whether he deferred to my appreciation or to his own will not be known, but our rapport was complete.

My father's work as a painter and decorator fostered his love for craftsmanship and gained him an enviable reputation. His services were in constant demand. However, in his spare time he could always be found pruning his shrubs out-of-doors or patiently attending his house plants. Our family's Sunday excursions into the meadows and swamps in search of unusual horticultural specimens seems far removed from the way present-day Sunday afternoons are spent, but that

FIGURE 163. In this nine-inch coral colored bloom of the *Haemanthus* each floret joins the next to create a geometric sphere.

segment of my childhood left an indelible impression on me and has been reflected in my later life.

My great aunt was a botanist. Her knowledge of Latin terminology, ecology, biology, and organic chemistry never ceased to awe and inspire me. She became another exemplar. There was seldom a summer day when we were not examining, analyzing, or tracking down, by means of the famous *Gray's Botany*, some rare specimen.

From my earliest days there has always been a garden. First, there were the more common annuals and perennials. These were followed by newer hybrids, which appeared in increasing numbers as man continued his experimentations in the plant world. Well-known plants and flowers became less challenging than the more exotic types, and those of foreign importation. Soil-conditioning, planting, transplanting, designing the garden, and developing specimen blooms still constitute my most creative activity.

What place has the above personal history in a book about enamel-

FIGURE 164. Splashings of steel have a kind of similarity to flowers.

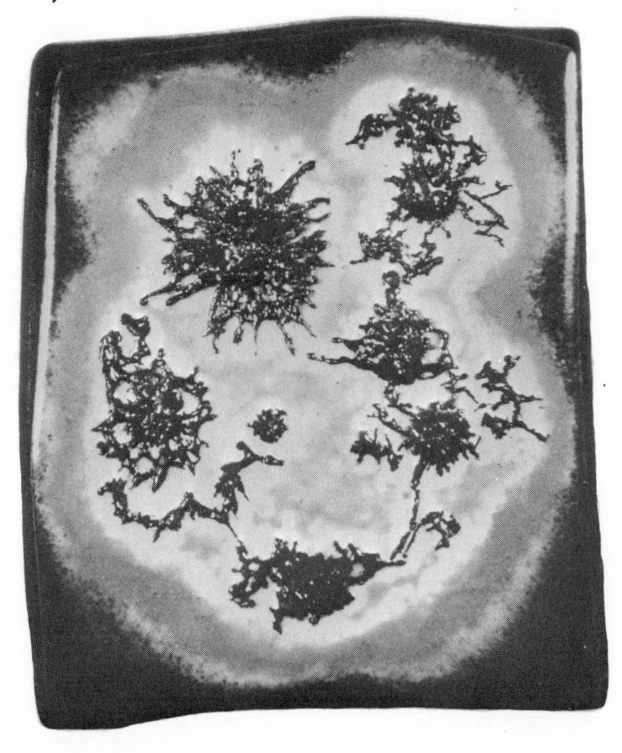

ing? It is inserted because I wish to write about flowers in particular as one source of inspiration for the enamelist. Almost every artist I know is instinctively enchanted with flowers and plants. But his interest is not that of the horticulturist, the university-bred botanist, nor even the professional flower arranger. He cares little about the garden encyclopedias, seldom reads flower magazines, is constantly confused by the seed catalogues, and rarely knows how deep to plant a bulb. His interest is visual—a visuality that involves more than seeing with his eyes. With him, it is more personal. His use of flowers in painting and applied design must be that which is first seen, then absorbed, then felt, and eventually interpreted through his own personality.

Flowers in History

Artists throughout history in their use of certain flower motifs have developed associations between those flowers and their countries—not because those flowers were the only ones available but because the artists felt an affinity for one flower rather than for another. For example, the lotus bud is associated with Egypt. In Medieval Europe both the carnation and the wild single rose (often shown with four petals) found their way as decoration in jewelry, textiles, furniture, and architecture. Heraldic trappings, tapestries, manuscripts, and paintings of French origin often employ a motif known as fleur-de-lis. (Our wild iris or "flag" is similar to the French fleur-de-lis, but the magnificent new hybrid irises of slightly different form had not been dreamed of in the thirteenth century when Charles V chose the design for the royal coat of arms.) We associate the chrysanthemum in its many curlicued manifestations with Chinese art, the laurel branch with classical Greece, flamboyant peonies with Japan, a twisted palm leaf with India, and so on.

Everyone thinks of tulips and Holland synonomously. A fascinating little book called *Tulipomania* written by Wilfrid Blunt tells about the wild speculation that went on regarding the propagating and selling of rare tulip bulbs in Holland during the second quarter of the seventeenth century. Royalty and connoisseurs of great wealth gambled their most cherished possessions and often their entire fortune for one unique bulb. To quote from *Tulipomania*: "The highest prices paid by tulip speculators bore no possible relation to the beauty of the flower. For one bulb, the following goods were given: 2 loads of wheat, 4 loads of rye, 4 fat oxen, 8 fat pigs, 12 fat sheep, 2 hogsheads of wine, 4 barrels of 8-florin beer, 2 barrels of butter, 1000 pounds of cheese, a

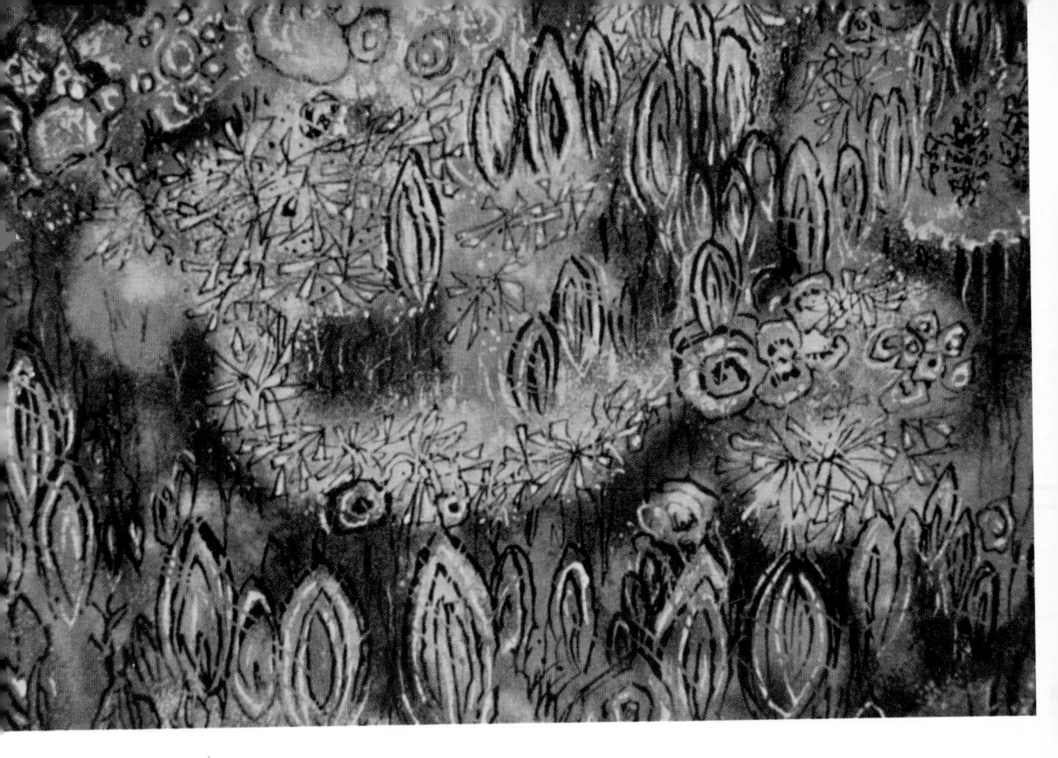

"MAY 15th" by the author; crocus, forsythia, and clear blue sky initiate the season with pastoral colors. (Courtesy Victoria Ball)

"OCTOBER 15th" by the author; ironweed and spent goldenrod stalks flash against autumn's vivid sky. (Courtesy George Gund Collection)

complete bed, a suit of clothes and a silver beaker—the whole valued at 2,500 florins." This story, which may seem a little ridiculous to us today, is not too far removed from what our modern hybridizers may secretly feel.

Flower Arranging

Beyond man's desire to grow flowers in order that he might feel they were "his," he has, since the twentieth century B.C., been so interested in his achievement that he has grouped several varieties together, brought them into the house, and displayed them to friends. He has exercised much imagination and ingenuity in the way he has done this. The Egyptians used flowers in stiff tiers; the Romans used baskets of roses and carnations, while the Byzantines liked tree-shaped arrangements. The symbolism of flowers was important to the Christians during the Gothic period; fruits and flowers were often combined by the Persians; and the use of cut flowers in bronze and pottery containers accented the Buddhist ritual in China.

In Japan, where flower arrangement has been an advanced study for centuries, we often find a stylistic grouping of a few simple flowers placed at the entrance of a home to express the host's message to his guests. French Rococo ornamentation in the eighteenth century included an abundance of bouquets of most elaborate description. Their porcelain figures, epergnes, and candelabra were massed with flowers of delicate color combinations. So also in the nineteenth century, when furniture and interior design reflected perfect proportions, the niches and mantels for arrangements of flowers in a classic tradition became austere and formal rather than free and casual.

Arranging and carefully combining flowers with appropriate containers has indeed become a most thriving form of expression in our time.

Flower Painting

Early flower paintings from Eastern civilizations (Chinese and Japanese dating from the seventh to seventeenth centuries) were based on a more philosophical conception than those of Western cultures. Eastern painters, with their poetic and contemplative point of view, considered flowers as a part of the great order of things in the universe. This was the same universe of which man was also but a small part.

Western painters were more apt to think of flowers as decoration, symbols, or offerings. They were depicted on canvas to appeal to man's finer feelings and nothing more. By the sixteenth century flowers had become subjects for paintings solely because of their own intrinsic beauty.

Chinese flower painting has had a long history. Most famous, perhaps, was Hsü Yüeh, who, in the early tenth century, became renowned as a painter of the lotus flower. At a later period it was Chao Chang who spoke of "the very soul of the flower." Such Japanese flower painters as Koyetsu (1557–1637) and Korin (1637–1716) were unique in combining rich colors with growth and compositional mastery.

Western flower painting can be traced from Egypt, where flowers were used as symbols, through Italian Renaissance, Spain, and eventually to the Netherlands. As mentioned before, the Dutch were instigators in tulip culture and undoubtedly rank with the English people as the great flower-lovers of all times.

Best known in the sixteenth and seventeenth centuries as masters in the art of flower painting were: Jan Bruegel (1568–1625) for his delicacy of texture, grace of growth, and close observation; Jan D. Van Heem (1606–1684) for his masterly technique; and lastly Jan Van Huysum (1682–1749) as the top master of the Flemish school. His combinations of colors, exquisite forms, and three-dimensional interpretations remain inspirational to this day, although sometimes the flower paintings of this era impress one as cold, technical, and perhaps lacking in the translucency or ethereal quality of flowers.

To a gardener, or an avid reader of seed catalogues, the most outstanding inconsistency of the Flemish flower paintings is that they were completely unseasonal. (One explanation is that these painters passed around illustrations of flowers from one artist to another—not unlike our present-day mailing lists—and merely compiled whichever flowers struck their fancy into one composition.) It has always concerned me, even as a boy, when viewing these famous still lifes how, if the artist were painting from nature (and how otherwise with such a display of horticultural exactitude), flowers of all seasons could be grouped together. In one picture there may be tulips, scillas, narcissi, and fritillaria of early spring; roses, peonies, and lilies of mid-season combined with chrysanthemums, apples, and grapes that mature only in autumn. Albeit, the tradition of the greatest flower painters of all time has never waned and never lost its impetus for later painters.

In France, Claude Monet (1840–1926) and Edouard Manet (1832–1883) were innovators when they showed colored flowers in brilliant surrounding light. Their rich arrangements of intense luminosity portrayed flowers in an entirely new kind of vibration.

Later Paul Cézanne (1839–1906) the great French post-Impressionist in his "Vase of Tulips" (1890) showed another kind of design by isolation of form so that each flower might be enjoyed separately. The enlarged close-ups of Georgia O'Keeffe (1887–), by deleting texture gave another consideration to the subject.

One might continue by mentioning Henri Rousseau (1844–1910) whose simple decorative patterns in "The Waterfall" (1910) established him as a recognized primitive; Odilon Redon (1840–1916) with his unearthly qualities and concentration upon the world of the mind; and Vincent Van Gogh (1853–1890) who with startling technical ability and turbulence portrayed flowers with a vitality never before witnessed.

Forms of Flowers for Artists

Unlike a special kind of genius such as Leonardo da Vinci, who was capable of excelling in many complicated fields of endeavor, the modern artist is fortunate if he finds time for only one or two of his interests. Economic pressure is such that a craftsman must spend most of his time in the studio. If he uses flowers as an inspiration in his work, it is unlikely that he would be an expert in the realm of botany or horticulture. He thinks and speaks of flowers in a kind of language familiar to his fellow artists, but not that used by the scientific researcher in the field of floriculture. Latin terminology, fine points in germination and growth, or generic classification mean very little to the basically visual person. It seems, then, that the artist is more concerned with shapes, forms, rhythms, structure, and color. As a flower begins to fade, it is obvious that it looses its commercial value. No florist would allow himself to sell a bouquet of tulips that had begun to droop. The same florist would be quite shocked, and no doubt unsympathetic, with a reputable artist who was visiting my home, and who, upon seeing some dark purple parrot tulips that had become dried and gnarled, spoke of their fascinating abstract beauty. Strange and exotic configurations of flowers after they twist and die, catching the eye of an artist, explain, in a sense, the point of view and the manner in which I wish to speak about flowers to the artist. As far as possible all Latin terminology will be supplanted with common names of flowers.

We shall consider flowers to be basically made up of radiating petals, spathes (sheaths) wrapped around a spadix (tongue), funnel shapes, umbels (umbrella-like forms), clusters or abstract informal shapes (orchids, bird-of-paradise blooms) etc. In that order they are bold, rounded, pointed, fine or fuzzy, massed, and linear. When using such material for an enameled bowl, plate, boxtop, panel, or mural the craftsman's concern for flowers resolves itself into vague but meaningful categories for him. The forms derived from his particular kind of flower knowledge are utilized only if they supply the dull, the static, the thin, the thick, the textured, the plain, etc., as the composition "grows." The painter when producing a still life of flowers also sees them in the above categories rather than as horticultural specimens. Even with a more naturalistic interpretation, he still instinctively takes the liberty of arranging the bouquet to comply with his design message rather than blindly imitating a photographic image.

Notwithstanding the fact that they have been historically and symbolically important, to the artist some flowers fall into a rather commonplace classification—for example, daisies, roses, tulips, and irises. A rose in full bloom, though popular with the layman, is not an inspiring design motif for the artist. It lacks form, shape, line, and pattern—much like a head of cabbage or a bunch of lettuce. Show an artist a lowly jack-in-the-pulpit, a French anemone, or perhaps the leaf from a philodendron plant, and his design thinking immediately becomes activated. The world of flowers abounds with endless strange and almost indescribable shapes and forms. The huge lobed leaves of the *Gunnera manicata* (forgive the botanical name, there is no common one) has umbrella-like leaves eight feet in diameter on prickly stems

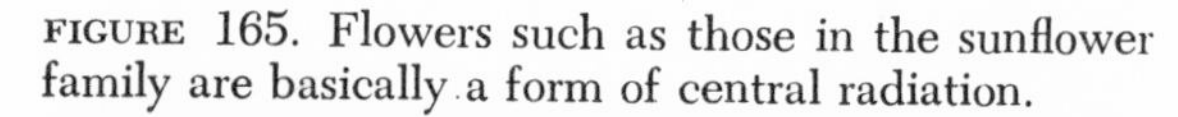
FIGURE 165. Flowers such as those in the sunflower family are basically a form of central radiation.

six feet high. The greenish flowers are clustered close to a majestic spike over three feet long. A flower known as "bottle-brush" has dense cylindrical flower spikes covered with coral-colored hairs that resemble its title. On a humid summer night one may ignite the volatile gas of the *Dictamus alba,* better known as "gas plant," creating a miniature fireworks display. Growing in the swamps of New England is a vine with insignificant small green blooms that gives off such a fetid odor that it has rightfully been named "carrion flower" (*Smilax herbacea*). For the spectacular-minded artist consider the *Allium giganteum* (giant flowering onion), which thrusts an enormous ten inch sphere composed of hundreds of tiny purple florets directly upwards on a straight five foot high tube. To complete such a list of flower fantasy would indeed take many volumes. The excitement of the flower world is to art what sound is to music.

Colors of Flowers

For the enamelist, certain phases of color are of vital importance. He may be inspired to transpose from a visual inspiration to his own

FIGURE 166. In this enameled box by the author the sunflower principle is dominant.

FIGURE 167. Clusters or umbels constitute a large segment of the flower world.

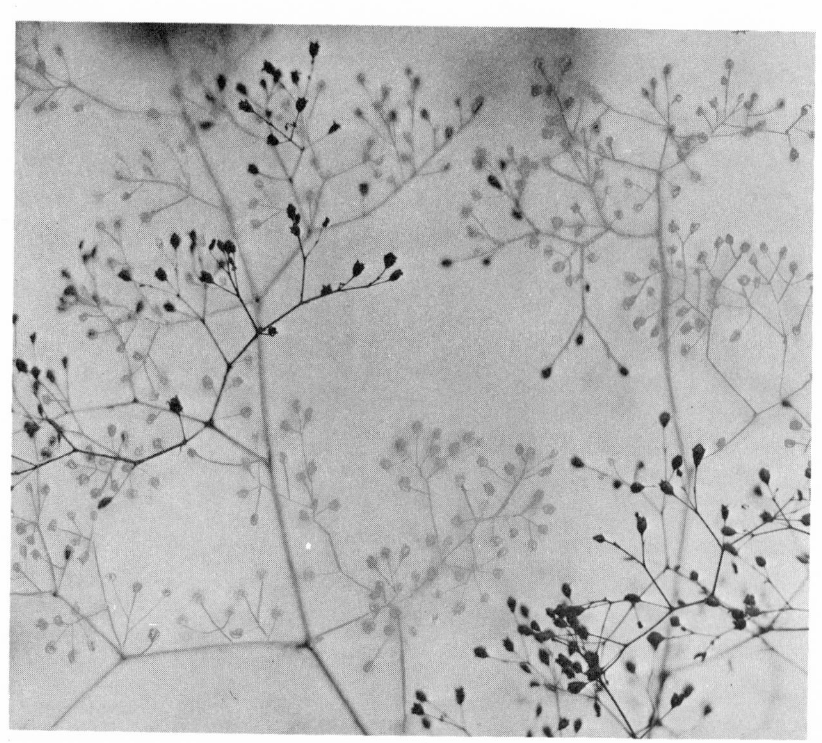

FIGURE 168. Seed pods of *Gypsophila* (Baby's Breath) suggest a design motif based on an umbrellalike form.

FIGURE 169. The obviously entitled bottle-brush flower is one of the most fantastic creations in the plant world.

FIGURE 170. The seed pods of the gas plant embody such design elements as variation of shape, size, placement, and rhythm. The pods look more like a spike of flowers than seeds.

medium such effects as deep water, textures of foliage, patterns of frost, etc., and from the study of flowers nearby every nuance of color can be found.

Iridescence, or that quality of color which gives a many-colored appearance when light rays strike the inner and outer layers of a subject, is of much concern to the enamelist. That quality is evident in many flowers. By cross-pollenating, plant hybridizers are creating new color combinations that test the imagination to describe or name. These names often result in such nondescript titles as "autumn-blend" for dahlias, or "copper tones" for a rose. Actually the fusing of various hues as a result of plant mutations invariably constitutes a kind of iridescence to that color. Velvet tones in gloxinias appear to be one color when seen from a given angle and another color in different lights. With many pure white lillies in the early morning, when the dew has not evaporated, there is a distinct iridescence to the petals that is not unlike mother-of-pearl. Place a sprig of wild jewelweed, which grows in the meadows, underneath a stream of fresh clear water. The result is an amazing iridescence turning each leaf into veritable silver foil.

Depth of color is of prime concern to the enamelist. By building up thin layers and fusing them separately he achieves the same quality that certain petals have under magnification. Lush mixtures of colors occur not only in the more evident blossoms of the plant world but also in stems (for instance a stalk of rhubarb). Crimson, scarlet, pink, white, and greenish white seem to flow together in fascinating admixtures. This is true of many flowers, stems, leaves, and branches. Greens, purples, and blues occur in endless fusings in delphiniums—brown and brick colors in the fascinating breeder tulips.

Inspiration for new and original surface treatments has no better source than the exquisite veining, striation, streaking, and barring that can be seen in the flower world. Study the striped surface of a jack-in-the-pulpit, the back side of a squash leaf, the endless variety of markings on dianthus pinks, petunias, and lemon lilies.

Some color accents are sharp and unexpected. Others of a more subtle nature occur at the tips of petals and are often hidden from view in the very center of the flower. Few who have never taken the time to look know that most red and cerise-colored tulips have at their centers an intense cobalt blue halo. This combined with jet black anthers and yellow pistils demonstrates one of nature's constant tricks—the color shock, the unexpected. Dissonances such as magenta red and scarlet vermilion often occur in the same gladiola bloom. Chartreuse is a color of recent development and one of unusual character. Growers are

FIGURE 171. The design principle of radiation is nowhere more magnificently embodied than in the nine-inch diameter seed pod of the *Allium giganteum*.

constantly striving for crossings that will produce a blue rose, a black orchid, or a pure white marigold. Such man-invented colors have commercial value, but I am not convinced that a blue rose would be more attractive to an artist than a common wild rose, regardless of the price.

Aesthetic experiences in the flower world exist for all to enjoy. There are complementary colors (red to green, yellow to purple) that may be observed in one small stem of the pestiferous dandelion or lowly clover. Poets speak of waving hordes of daffodils, and aspen leaves shivering in the breeze, or of a rose being a rose. Such sentimental interpretations or even humanization of flowers is to the artist more appealing than the search for deviations and variations.

FIGURE 172 and 173. With the close-up lens the camera reveals such hidden secrets as these tri-pocketed seed capsules.

Sculpture in Flowers

Although the enamelist is primarily concerned with two-dimensional interpretation, his inspiration for that interpretation is often derived from three-dimensional manifestations in nature. In welded forms and in forged or hammered pieces of jewelry, both of which might receive enamel, a conception of sculptural organization is of prime importance. Carl Fabergé at the turn of the century (page 30) produced numerous examples of realistic flowers enameled in a method called encrusté (thin layers of enamel fired over a modeled surface). Today Swiss and Swedish craftsmen manufacture flower forms enameled on gold and silver to be used as brooches or earrings.

Flowers present themselves as a point of departure for the designer, rather than as a limit on the imagination. Certain flowers are of such sculptural significance—calla lily, Mexican torch lily (Tigridia), or the hibiscus family—that a stylistic interpretation is almost obvious. Any flower in the large group that includes, arums, skunk cabbage, jack-in-the-pulpit, anthuriums, and callas have as their most noticeable characteristic sculptural potentialities. One might say that the three-dimensional aspect of a flower like the passion-vine bloom or the sunflower suggests a kind of conventional treatment. Witness the

FIGURE 174. The *Datura* is a flower which presents a strong and simple sculptural statement.

conventional spiral of a fern frond, or unfolding rose bud. Certainly, the seed pod of the dandelion, or the more spectacular *Allium giganteum* is geometric. Sculptors, as well as enamelists, have been preoccupied with this "explosive manifestation" of the flower world for some time. Radiation from a point (a well-known principle in basic design) is constantly seen in nature.

One sees in flowers and plants many types of growth so amorphous and abstract that they defy description. One of the huge informal decorative dahlias that I have grown, known as Kiyo Hime, a Japanese importation (reaching a size of 13″ diameter by 8″ in depth) is so fantastically irregular, and so unpredictably splashed with crimson on a yellow ground, that it appears more like a nonobjective painting (Fig. 179).

So it would seem that, whether an artist is attracted to some undefined free expression found in the way clematis blooms occur on a trellis or to an unerring mechanical perfection such as the placement of seeds in a sunflower, flowers are an inexhaustible source of delight and inspiration.

FIGURE 175. *Amorphophallus* is a large sculptural plant with liver-colored blooms.

FIGURE 176. "Amorphophallus," detail of panel by the author showing overglaze application as a final technique.

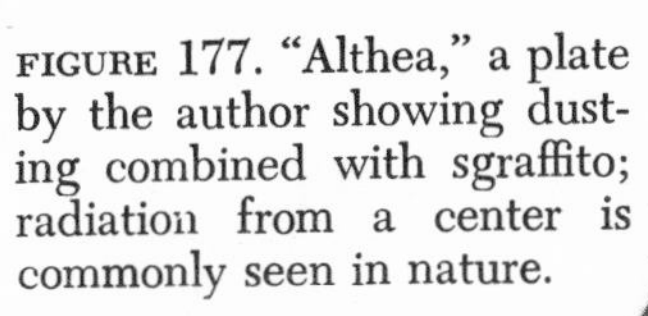

FIGURE 177. "Althea," a plate by the author showing dusting combined with sgraffito; radiation from a center is commonly seen in nature.

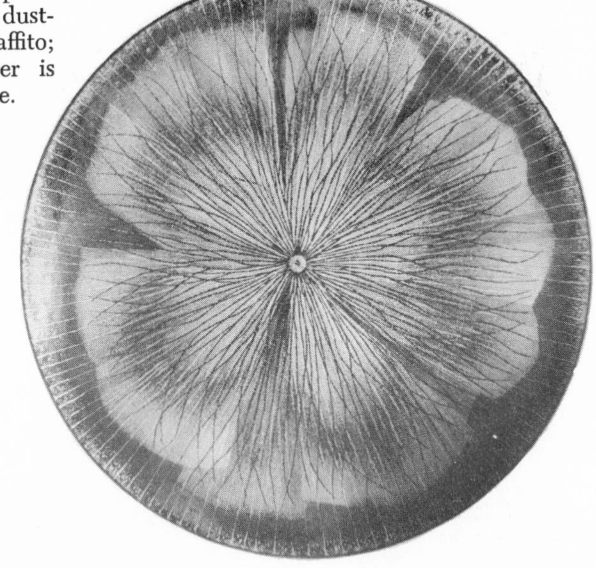

FIGURE 178. The six petals of the lily family radiate from a given point.

FIGURE 179. The Japanese informal decorative dahlia has blooms 13 inches across and 8 inches in depth.

Seed Pods

Seeds are to be considered with awe and amazement. Some exist in microscopic sacks, the seed themselves being finer than dust. On the other hand, think of the ornamental catalpa tree with its clusters of orchid-like blooms and seed pods over 18 inches long and of the huge egg-like seed of the avocado. Seeds with their endless variation of shapes, sizes, and colors are, in fact, more fantastic than flowers.

Some seeds look exactly like enlarged spiders or insects, a device cunningly perpetrated by nature to deceive birds. Such seeds having been mistakenly devoured as insects by birds are eventually located far from the mother plant by means of the bird's excreta. Some seeds have wings or fuzz or parachutes for the similar purpose of dissemination to new localities. These might be called the kinetic or mobile group of seeds. The tiger lily forms small black bulblets at each bract. These, when released, roll along an accompanying grooved leaf, thus assuring their placement away from the original plant. Some seeds

FIGURE 180. The brilliant orange seed pods of the Japanese lantern suggest usable design shapes.

FIGURE 181. Sculptural forms seen in the scales of the artichoke seed pod.

FIGURE 182. The Egyptian lotus seed head is handsomely sculptural; notice how each oddly shaped pocket contains one hard seed.

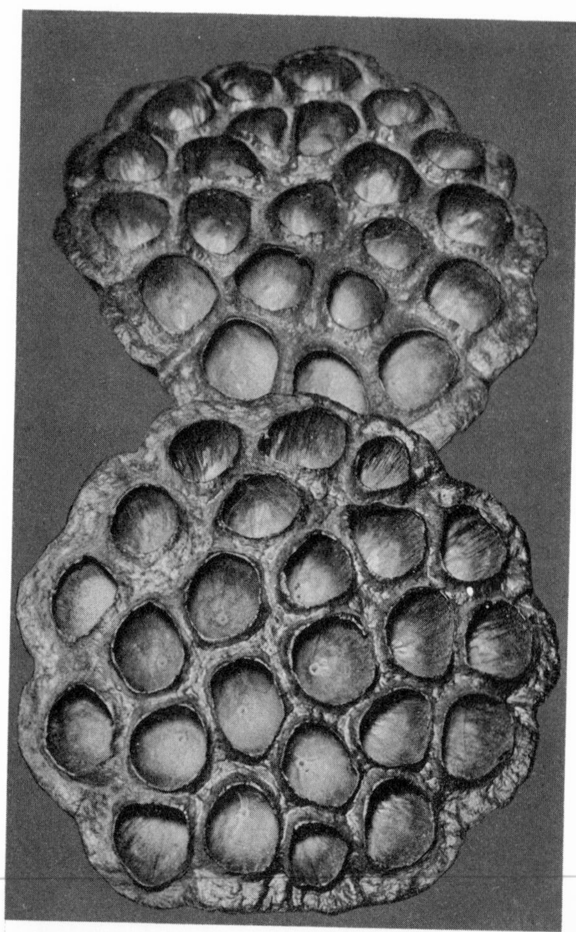

are exploded (jewelweed), some catapulted (cleome), some take to the air on perfectly constructed wings (sugar maple). All berries, which are seeds, are part of this contrived natural plan to use birds as carriers.

The miracle of the watermelon—a huge fruit, pink, white, green, sweet, and full of liquid—that is produced from a small black seed (or the fact that a pumpkin weighing over forty-five pounds can be grown from an insignificant seed) is philosophical or religious in aspect —an aspect related to the artist's sense of wonderment, that quality of mind which prefaces imagination.

The shape of the poppy seed pod is not unlike an elegant knob or finial to a covered enamel bowl. Vase shapes and pottery shapes considered as most original conceptions are often but reflections of fascinating seeds or seed pods.

A most unusual seed pod, and one that offers decorative possibilities for the enamelist, is honesty (sometimes called dollar-plant, moonwort, satin-flower, St. Peter's penny, or moneyplant). The flower, which comes in the spring, is a magenta phlox-like bloom of no special significance, but when the seeds are formed, they are encased in a round white translucent membrane about the size of a silver dollar. This and endless other seed pods—smooth, hard, spiney, burr-like,

FIGURE 183. How simply and yet how magnificently the Oriental poppy seed head suggests the filial to a covered bowl.

FIGURE 184. "Pomegranate," a plate by the author showing sgraffito light line and overglaze dark line. (*The Art Gallery of the University of Notre Dame*)

FIGURE 185. Delightful freedom and style may be observed in the silvery seed pods of "honesty," or dollar plant.

fluffy, or hair-like—serve as constant inspiration for the designer-craftsman.

Foliage and Textures

In enameling, textures are usually the result of technical experiments, those peculiar to the medium itself or as a consequence of fusability. However, there exist many sources of inspiration for planned textures, one of which is the world of foliage. Thinking of foliage as multiple grouping of leaf shapes, juxtaposed, separated, or overlapping, one is able to devise new surface patterns. Applying some concern for the negative and positive areas, one should experiment with foliage textures using various sgraffito techniques. (Tests for sgraffito, page 74.)

Study the subtle variations of the pine (*pinus*) family. The unusual geometric surface tensions will add to an enamelist's repertoire of motif material. There is a vast variety of pines including: white pine, with its widespread branches and gray-green needles; red pine, with glossy green needles; Japanese black pine, having bright green, sharp-pointed needles and sometimes needles banded with two yellow stripes; Austrian pine, with stiff, dark green needles; Japanese red pine, whose

FIGURE 186. "Honesty" by the author; inspired by the plant of the same name.

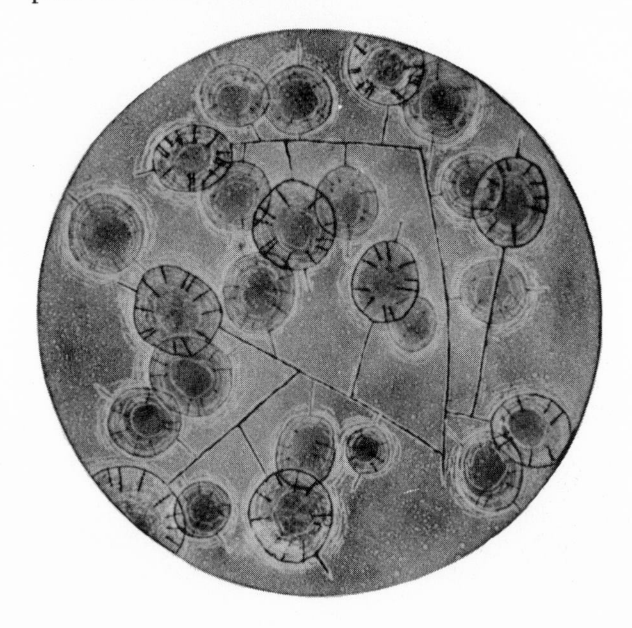

FIGURE 187. Ralph Marshall has presented here a dramatic design of a single leaf with its inner structure.

needles are slender and blue-green; and Scotch pine, having needles that are often twisted. The variations continue in the silhouette of the trees, the kinds and colors of needles, and the shape and construction of the cones.

One of the most fascinating trees I have ever seen is the monkey puzzle tree. The odd shaped, sharp-pointed leaves clothing branches and trunk make extraordinary texture patterns. They are so constructed and interwoven that they present a problem for climbing, even for a monkey (thereby its common name).

From the exotic to the conforming, foliage textures present vivid contrasts. There are lamb's ears with soft hairy down covering them; crossandra leaves, which are so shiny and polished they appear to be oiled; squash leaves, which are rough or bristly; warty rhubarb leaves. The list is neverending. Especially when designing a flower garden, the interplay of foliage, flat leaves, linear, fine cut, vertical, horizontal, twisted, and simple must be employed for "color" as well as blooms.

Maltreatment of Flowers in Design

The maltreatment of flowers in design, which is seen all too often in exhibitions including enameling, comes mainly from a lack of knowl-

FIGURE 189. The textures found on the back side of an ordinary summer squash are fascinating to the designer.

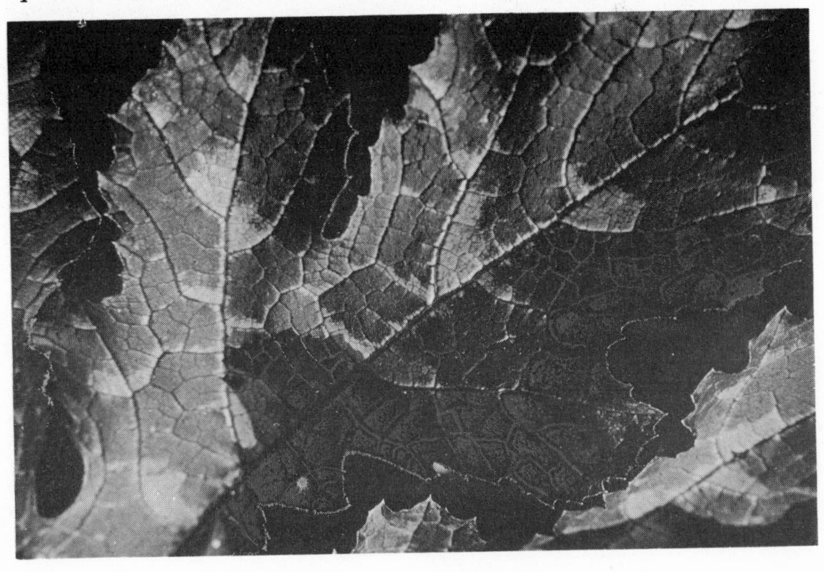

edge about how things grow. Such knowledge is sometimes instinctive with those who live or have been brought up in the country, but the city dweller need not be oblivious to the laws of growth and structure. More keen observation and some appreciation would improve the caliber of the hobbyist's work, particularly when he concerns himself with the precarious and constantly misused flower motif.

At the risk of being called both pedantic and conservative, I shall list, in the way of criticism, the most frequent abuses to flowers when used as a motif in design.

First, the most basic of all faults is the lack of structure. In the most abstract conceptions there is always structure in the form of an unseen compilation of axes, median, stresses, and the framework upon which the composition is built. More essential, perhaps, is the need of such framework or basic structural armatures when flowers are interpreted literally. To relate a stem to a bloom (a line to a disk) in order to have it balance on the center point as it does in most flowers, one must draw it that way. If in perspective views the continuation of the stem is not seen, it must give the "appearance" of follow-through if it is to have structural evaluation.

Lack of structure and lack of growth are not quite synonomous. In certain Italian Renaissance scrolls the artists have taken liberties with principles of growth. This liberty has also been taken in illuminated

FIGURE 190. The monkey puzzle tree is as intriguing aesthetically as it is to a monkey who finds himself hopelessly entangled in its branches.

manuscripts. We are speaking of vines or tendrils, for instance, where the leaves appear to be growing off the stem in the wrong direction. No doubt, as an expedient for space filling this kind of designer's liberty was taken. However, with no concern for growth principles, the design can become very unpleasant and awkward. One expects a flower, a plant, a tree to grow upward and outward. To progress in the opposite direction is distressing unless done with special intent. Aesthetically or logically (two words not always similar in meaning) inner growth can be "sensed" even if it is not always obvious. (We are speaking of designing with flowers, not the rendering of botanical plates with anatomical exactitude.) There is a subtle difference between freedom of expression and the utter lack of knowledge or perception as to how a plant grows.

Dwarfing a motif or forcing it into a shape that is too small for it is unpleasant. A sense of scale is not something that can be easily taught by an instructor, or, if it is taught in the classroom, the student will still have to make his own decisions regarding scale, relying only on his inherent sense of proportions. The novice gardener is confused by scale, as most seed catalogues picture a 12″ diameter dahlia bloom the same size as a 2″ diameter petunia. Of course, an artist need not be that logical when producing an enameled panel, but note, for example, that the flower forms chosen by Georgia O'Keeffe for her oversized interpretations are those that accept her particular kind of stylization "scalewise."

The argument of whether or not any flower forms should be placed on any preconceived shape is a moot one. Perhaps it would be best to remain negative about the whole subject. Certainly, one sees too many enameled plates and bowls that would have been better left "undecorated." The worst enamels are those upon which a handful of unrelated flower shapes seem to have been "dropped." *Only* if flower forms can be so abstracted or integrated to represent a textural surface should they be employed for this medium. It is a most distasteful use of flowers to paint them realistically on salad bowls, salad forks and spoons, clocks, dishes, handbags, rugs, towels, and lampshades. In desperation, the shopper with taste wonders why flowers must be "stuck" on everything? Better that the flowers should be left growing in the garden!

Another criticism and one that is rather complex is the poor choice of color for flowers. This lack of feeling for color is fostered considerably by the flourishing trade in artificial flowers. Most florists display, along with their living flowers, an abundance of plastic imitations. Some are rather convincing, all are indestructible, most are washable, but color-

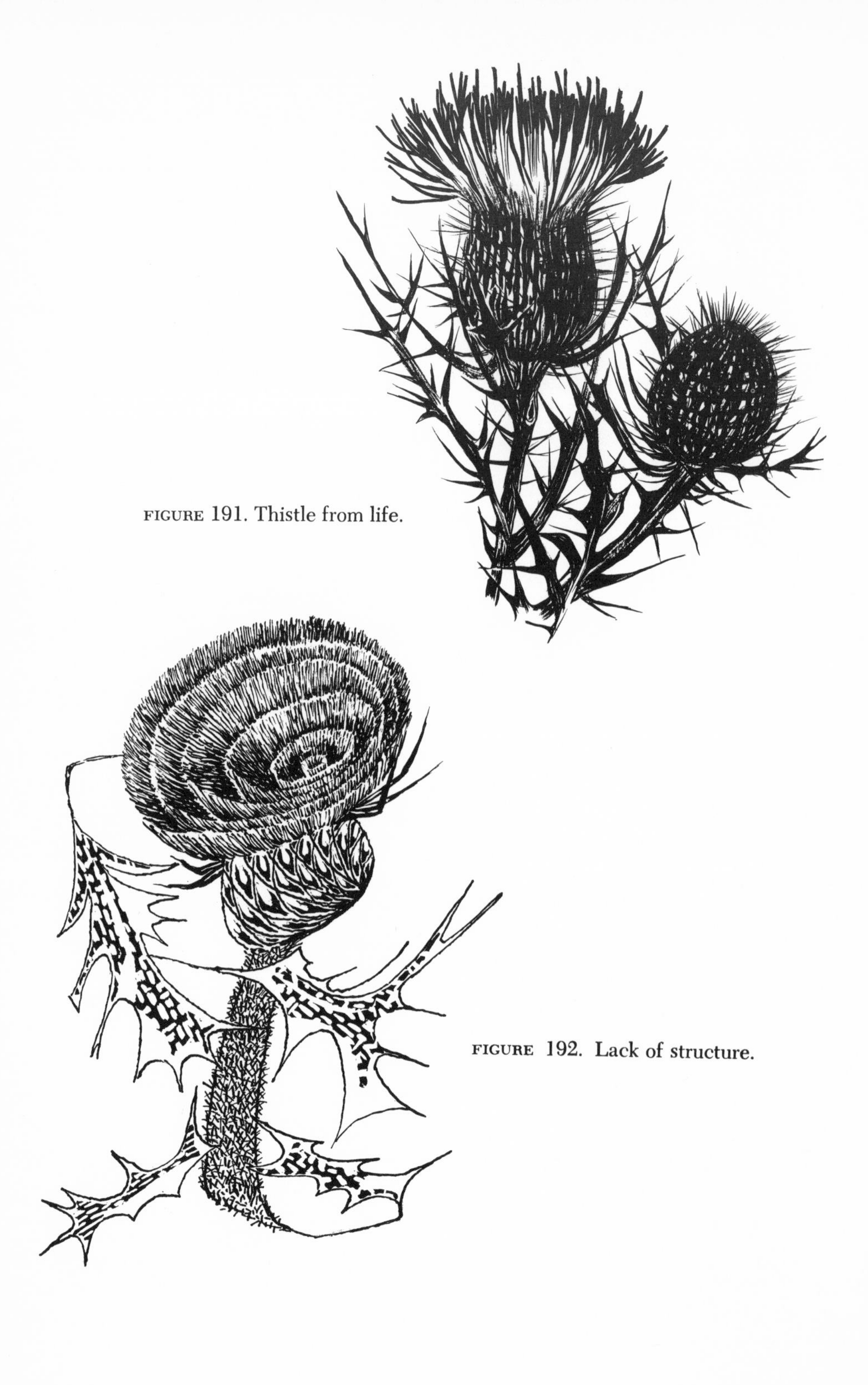

FIGURE 191. Thistle from life.

FIGURE 192. Lack of structure.

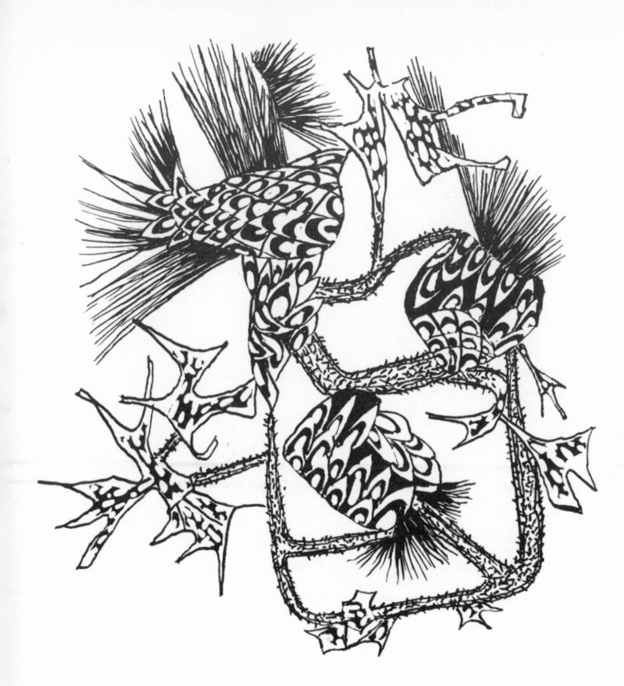

FIGURE 193. Lack of growth.

FIGURE 194. Lack of scale.

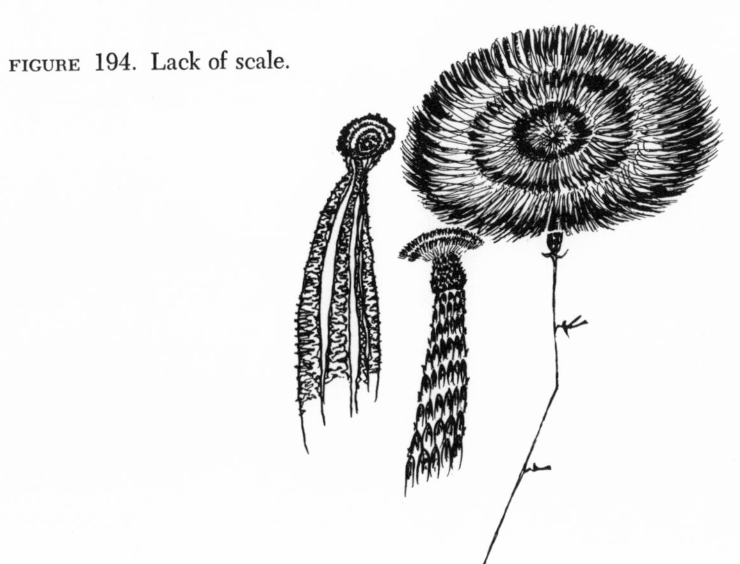

wise many are outrageous. I have seen artificial snapdragons in vitriolic blue-green, oversized wax roses a brilliant magenta, and huge orange water lilies. This kind of unnatural coloration galls the artist, the gardener, and the genuine flower lover. One sometimes sees the same crude and distasteful maltreatment of flower colors in the work of the dilletante artist.

FIGURE 195. Lack of appropriateness to shape.

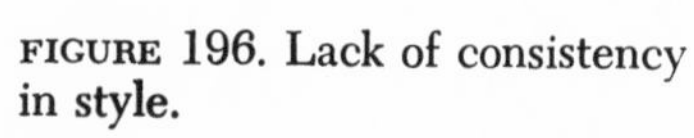

FIGURE 196. Lack of consistency in style.

FIGURE 197. Lack of contemporary feeling.

FIGURE 198. Lack of drawing.

FIGURE 199. Trite motifs.

One certain way to have a design suffer from lack of style is to allow it to be awkward or clumsy. Stems that are too thick with flowers too shrunken, fat, uninteresting leaves, poorly chosen subject matter, dull and uninteresting color, a heavy, thick technical handling—all contribute to lack of style. With whatever style is intended (geometric, abstract, stylistic, or impressionistic) the artist must be consistent as the design develops. Most important, each individual must attempt to achieve his own sense of style.

One might say that bad structure, growth, and scale constitute lack of drawing, as, of course, they do, but drawing implies even more. It also involves perception. The act of drawing is more than observation and certainly more than acquired skill. It is an act of intuition and imagination. An interpretation of flowers for a circular, rectangular, or square utilitarian object demands more intuition about flowers than does "rendering" a still life bouquet. Growth, structure, scale, *and* intuitive drawing must synchronize to make the flowers seem to exist, regardless of how abstract the interpretation.

Considering the fantastic realm of the flower kingdom, the wonderment and excitement of the natural phenomena, one wonders why some commercial manufacturers persist in merchandising handkerchiefs with trite daisy designs in the corners, towels with borders of classical Greek leaves, printed textiles using sentimental violets, wild roses, and forget-me-nots, and birthday cards embossed with clinging ivy, and other well-known triviality. The designer might benefit by turning the pages of a good seed catalogue, or better still, if he is a confined urban-dweller, visit a conservatory. Triteness implies taste or, rather, lack of taste. Poor taste is hard to define, and a dangerous term for the critic or the art instructor to use. (Paradoxically many untutored people have an uncanny sense of taste.) Taste can only be improved (if at all) by association, by study, and by constant observation of current trends and exhibitions. Richard Gump in his book *Good Taste Costs No More* (Doubleday, New York, 1951) states: "There is no doubt that the level of taste is steadily rising, that more and more people are becoming conscious of the nature and value of good design." Let us hope that his statement is not too optimistic and that enamelists who use flowers as a design motif will show more cognizance and appreciation of the world around them.

Conclusion

An enamelist should never lose track of, nor forfeit, the excitement he experienced when he took the first piece from the kiln. He must never lose his conviction that enameling is in its infancy and that his is the opportunity to carry it further. As anyone who claims to be creative recognizes, a craftsman is intermittent, seasonal, perhaps cyclic in his productivity. Artistic expression does not flow from anyone in a steady stream. For this reason, it seems to me that taking care of a potted plant, observing a seed germinate, or watching a bud unfold—being both creative and prepossessing—supplies the perfect instrument for the artist's discontinuousness. His good fortune is that his work is never finished; he is so preoccupied that youth and old age blend unnoticeably. Similarly, nature is never static—never conclusive, never definitive. The parallelism between enameling and growing things is clear: neither has revealed its infancy and neither has revealed its eventuality.

FIGURE 200. The author's garden on Lake Erie.

BIBLIOGRAPHY

Addison, Julia DeWolf. *Arts and Crafts of the Middle Ages*. Boston: L. C. Page, 1908.

Albers, Josef. *American Abstract Artists*. New York: The Ram Press, 1946.

Austrian Applied Arts. Wien: Verlag Heinz & Co., 1930.

Bager, Bartel. *Naturen Som Formgivare*. Stockholm: Nordisk-Rotogravyr, 1957.

Bainbridge, Henry C. *Peter Carl Fabergé*. London and New York: B. T. Batsford, Ltd., 1949.

Ball, Victoria Kloss. *The Art of Interior Design*. New York: MacMillan, 1960.

Bates, Kenneth F. *Basic Design, Principles and Practice*. New York and Cleveland: World Publishing Co., 1960.

———*Enameling, Principles and Practice*. New York and Cleveland: World Publishing Co., 1951.

Blunt, Wilfrid. *Tulipomania*. Harmondsworth, Middlesex, England: Penguin Books Ltd., 1950.

Britton, N. L. and Brown, A. *Illustrated Flora*. Vols. I, II, III. New York: Scribner's, 1913.

Brown, W. N. *The Art of Enameling*. London: Scott, Greenwood & Co., 1900.

Bulletin, The. The Cleveland Museum of Art. Vol. LIII, No. 3, Mar. 1966.

Burger, Willy. *Abendländische Schmelzarbeiten*. Berlin: R. C. Schmidt and Co., 1930.

Burris-Meyers, Elizabeth. *Historic Guide to Color*. New York: W. Helburn Inc., 1951.

Carson, Rachel. *Silent Spring*. Boston: Houghton Mifflin Co., 1962.

Cunynghame, H. H. *Art Enameling on Metal.* Westminster, England: Archibald Constable and Co., 1899.

Dawson, Mrs. Nelson, *Enamels.* (Little Books on Art Series). Chicago: A. C. McCling & Co., 1908.

De Koningh, H. *The Preparation of Precious and Other Metal Work for Enameling.* New York: The Norman W. Henley Publishing Co., 1930.

Dobkin, Alexander. *Principles of Figure Drawing.* Cleveland and New York: World Publishing Co., 1948.

Emerson, Sybil. *Design, A Creative Approach.* Scranton: International Textbook Co., 1953.

Enamels. Museum of Contemporary Crafts. New York: Clark and Way, 1959.

Encyclopaedia Britannica, "Enamel." Vol. 8.

Encyclopedia of World Art. Vol. IV. New York: McGraw-Hill, 1961.

Evans, Joan. *A History of Jewelry 1100–1870.* London: Faber & Faber, 1953.

Everett, T. H. *New Illustrated Encyclopedia of Gardening.* New York: Greystone Press, 1964.

Feirer, John L. *Modern Metalcraft.* Peoria, Illinois: The Manual Arts Press, 1946.

Fisher, Alexander. *The Art of Enameling Upon Metal.* London: The Studio, 1906.

Gardner, Helen. *Art Through the Ages.* New York: Harcourt, 1959.

Gump, Richard. *Good Taste Costs No More.* New York: Doubleday, 1951.

Hildburgh, W. L. *Medieval Spanish Enamels.* London: Oxford University Press, 1906.

History of Enamel VII-XX Century. Dept. of Fine Arts. Univ. of Pittsburgh, 1950.

House, Homer D. *Wild Flowers.* New York: MacMillan, 1935.

Jessup, Ronald. *Anglo-Saxon Jewelry.* New York: Hilary House, 1950.

Kassler, Elizabeth B. *Modern Gardens and the Landscape.* New York: The Museum of Modern Art, distributed by Doubleday and Co., Inc., Garden City, New York, 1964.

Kepes, Gyorgy. *Language of Vision.* Chicago: Paul Theobald, 1941.

Krutch, Joseph Wood. *The Gardener's World.* New York: Putnam's, 1959.

Labarte, Jules. *Peinture en Émail—Recherches sur La Peinture en Émail dans L'antiguité et au Moyen Age.* Paris: Victor Didron, 1856.

Lavedan, Pierre. *Léonard Limosin et Les Emailleurs Français.* Paris: Henri Laurens, 1913.

Lehnert, Georg Hermann, ed. *Ilustrierte Geschichte des Kunstgewerbes,* Vol. I. Berlin: M. Oldenbourge, 1907.

Lenning, Henry F. *The Art Nouveau.* The Hague: Martinus Nyhoff, 1951.

Marcus, Margaret Fairbanks. *Period Flower Arrangement.* New York: M. Barrows, 1952.

Maryon, Herbert. *Metalwork and Enameling.* New York: Watson Guptill Publications, 1958.

Matisse, Henri. *Jazz.* Munich: R. Piper and Co., 1947.

Meyers, Francis J. *Charcoal Drawing.* New York: Pitman, 1964.

Milliken, William M. "Early Enamels in the Cleveland Museum of Art." *Connoiseur,* Vol. 76, pp. 66–74, Oct. 1926.

Morgan, J. Pierpont. *Catalogue of the Collection of Jewels and Precious Works of Art.* London: Chiswick Press, 1910.

Neuburger, Albert. *The Technical Arts and Sciences of the Ancients.* New York: MacMillan Co., 1930.

Oshikawa, Josui and Gorham, Haze H. *Manual of Japanese Flower Arrangement.* Japan: Nippon Bunka Renmei, 1936.

Pack, Greta. *Jewelry and Enameling.* New York: D. Van Nostrand, 1941.

Rogers, Frances and Beard, Alice. *5000 Years of Gems and Jewelry.* New York: Frederick A. Stokes Co., 1940.

Rupin, Ernest. *L'oeuvre de Limoges.* Paris: Alphonse Ricard, 1890.

Seymour, E. L. D. *The Garden Encyclopedia.* New York: Wm. H. Wise Co., 1936.

Shahn, Ben. *The Shape of Content.* Cambridge, Mass.: Harvard University Press, 1957.

Sitwell, Sacheverell. *Fabergé, An Exhibition at a la Vieille Russie.* New York: A La Vieille Russie, 1949.

Snowman, Kenneth. *The Art of Carl Fabergé.* Boston: Boston Books, 1962.

Stechow, Wolfgang. "Ambrosia Bosschart, Still Life," *The Bulletin of the Cleveland Museum of Art.* C. M. A. Press, Vol. LIII, No. 3, Mar. 1966.

Thompson, Thomas E. *Enameling on Copper and Other Metals.* Highland Park, Illinois: T. C. Thompson Co., 1950.

Tucker, Allen. *Design and the Idea.* New York: Oxford, 1939.

Untracht, Oppi. *Enameling on Metal.* Philadelphia and New York: Chilton Co., 1957.

Wiener, Louis. *Handmade Jewelry.* New York: D. Van Nostrand, 1948.

Winter, Edward. *Enamel Art on Metals.* New York: Watson Guptill Publications, 1958.

INDEX

ABOUT THE AUTHOR

Kenneth F. Bates, designer, craftsman, author, teacher, and lecturer. His best-selling book, *Enameling: Principles and Practice,* was published in 1951, and *Basic Design: Principles and Practice* in 1960. His work in enameling is known nationally and internationally. His numerous professional prizes in art are supplemented by a bronze medal for horticultural achievement and a silver medal from the American Rose Society. He won the Horace Potter Silver Medal for Excellence in Craftsmanship in 1949, 1957 and 1966. In 1959 his work was exhibited in the Brussels World Fair, and he was a major contributor to a retrospective exhibition of American enameling at the Museum of Contemporary Crafts in New York. He has since had one-man shows in Pittsburgh, Chicago, Cleveland, Brooklyn, and the University of Notre Dame. In 1963 he received Cleveland's top honor, the Visual Arts Award. In 1965 he was given the first Faculty Grant from the Cleveland Institute of Art, which enabled him to study enameling and horticulture in Spain, France, Holland, England, Scotland, and Wales. He has been an instructor in design and enameling in that Institute for 40 years.

A BRIEF HISTORICAL OUTLINE

	8th c. A.D.	9th c. A.D.	10th c. A.D.	11th c. A.D.	12th c. A.D.	13th c. A.
GREEK						
CELTIC AND ROMAN		Last of Celtic enamels found in Ireland and England				
EGYPTIAN						
JAPANESE	Primitive mirror with enamel found	Naturalistic birds and flowers, cloisonné enamel	Work exported in large quantities	Cloisonné enameling continued until present time		
ANGLO-SAXON		The "Alfred" jewel dated 849 – 901	Later Saxon period, gold enamel and granulation			
INDIAN (EARLY PERIOD)						
INDIAN (LATER PERIOD) PERSIAN						
BYZANTINE	Rare enamels brought back from Crusades		Peak of Byzantine enamels	Pala d'oro now at St. Marks in Venice		
FAR EASTERN CHINESE						
GERMAN				Important Guelph Treasure, 1040	Mosan School cloisonné and champlevé	
MIDDLE AGES EUROPEAN				English necklace reflects Byzantine School		Gold pendan for Edward I dated 1290
GOTHIC PERIOD EUROPEAN						
RENAISSANCE ITALIAN						
FRENCH LIMOGES SCHOOL						
SWISS						
ENGLISH						
RUSSIAN (FABERGÉ)						
AUSTRIAN						
AMERICAN						